Synchronization in Wireless Sensor Networks: Parameter Estimation, Performance Benchmarks and Protocols

Erchin SERPEDIN

Texas A&M University

AND

Qasim M. CHAUDHARI

Iqra University

CAMBRIDGE UNIVERSITY PRESS

Cambridge, New York, Melbourne, Madrid, Cape Town, Singapore, São Paulo, Delhi

Cambridge University Press

The Edinburgh Building, Cambridge CB2 8RU, UK

Published in the United States of America by Cambridge University Press, New York

www.cambridge.org

Information on this title: www.cambridge.org/9780521764421

First published 2009

Printed in the United Kingdom at the University Press, Cambridge

A catalogue record for this publication is available from the British Library

Library of Congress Cataloging in Publication data

Serpedin, Erchin, 1967–

Synchronization in wireless sensor networks : parameter estimation, performance benchmarks, and protocols / Erchin Serpedin and Qasim M. Chaudhari.

p. cm.

Includes bibliographical references and index.

ISBN 978-0-521-76442-1 (hardback)

1. Wireless sensor networks. 2. Wireless sensor networks. 3. Synchronous data transmission systems. 4. Timing circuits. I. Chaudhari, Qasim M. II. Title.

TK7872.D48S47 2009

681′.2 – dc22 2009009344

ISBN 978-0-521-76442-1 hardback

Synchronization in Wireless Sensor Networks: Parameter Estimation, Performance Benchmarks and Protocols

Wireless sensor networks are set to play a key role in a wide range of civilian and military applications, with tiny sensors connected through wireless links performing various sensing, computing, communication, and control tasks in highly distributed systems. This book presents a critical element in the deployment of wireless sensor networks: the process of synchronization. It summarizes the most important clock synchronization protocols proposed for wireless sensor networks, with special emphasis placed on deriving efficient clock offset estimation schemes and performance benchmarks. Graduate students of electrical and computer engineering and computer science will find this a valuable resource, as will engineers who are interested in designing efficient clock synchronization algorithms and improving the performance of existing synchronization protocols.

Erchin Serpedin is currently an Associate Professor in the Wireless Communications Laboratory at Texas A&M University, where he joined after receiving his Ph.D. in Electrical Engineering from the University of Virginia, Charlottesville, in 1999. His research interests lie in the areas of statistical signal processing and wireless communications. Dr. Serpedin has served as Associate Editor for numerous journals including *IEEE Transactions on Wireless Communications*, *IEEE Transactions on Signal Processing*, *IEEE Transactions on Communications*, *IEEE Signal Processing Letters*, and *IEEE Communications Letters*.

Qasim M. Chaudhari was awarded his Ph.D. in Electrical Engineering from Texas A&M University in 2008 and is currently an Assistant Professor at Iqra University, Islamabad, Pakistan. Before entering academia, he worked with the SoC Tools Group of Communications Enabling Technologies, Islamabad, and later with the HSDPA performance test team of Qualcomm Inc., San Diego. His research interests include digital communications, estimation and detection theory in general and channel estimation and synchronization in wireless sensor networks in particular.

To our parents and families,
Zara, Nisa, Aisha, Nesrin and Maryam.

CONTENTS

PREFACE

The clock or time synchronization problem in wireless sensor networks (WSNs) requires a procedure for providing a common notion of time across the nodes of WSNs. In general, clock synchronization is viewed as a critical factor in maintaining the good functioning of WSNs due mainly to their decentralized organization and timing uncertainties caused by the imperfections in hardware oscillators and message delays at the physical and medium access control (MAC) layers. In addition, synchronization of the nodes of wireless sensor networks is crucial for implementing fundamental operations such as power management, transmission scheduling, data fusion, localization and tracking, and security protocols to name only a few applications.

The aim of this book is to provide an introduction to the clock synchronization problem of WSNs from a statistical signal processing viewpoint. Therefore, most of the topics presented in this book deal with building efficient clock offset estimation algorithms and performance benchmarks for general synchronization approaches that rely on sender–receiver and receiver–receiver timing packet exchange mechanisms. A summary of the key features of the most representative protocols proposed for clock synchronization of WSNs is also presented, together with some interesting open research problems.

Synchronization of WSNs is currently a very active research field with a large number of results and very diverse contributions coming from an equally diverse body of researchers: computer scientists, electrical engineers, mathematicians, statisticians, etc. Despite the deployed efforts, the general problem of building efficient global synchronization protocols for large-scale wireless sensor networks is still open and the proposed results are still introduced in a quite ad-hoc manner, lacking comprehensive design and optimization studies to assess and improve their performance in a systematic fashion. Although herein we will not solve these general and very important problems, this book assumes the modest task of building efficient synchronization algorithms and performance benchmarks for simple sender–receiver and receiver–receiver timing packet exchange based synchronization protocols. Despite its unambitious goal, this book will be of interest to all practitioners who are looking for techniques to improve the performance of existing protocols such as network time protocol (NTP), time protocol for synchronization of sensor networks (TPSN), reference broadcast synchronization (RBS), pairwise broadcast synchronization (PBS), etc., and to develop novel synchronization protocols.

This book consists of 15 chapters. Chapter 1 is a short introduction to the time synchronization problem, its history, and importance. Chapter 2 presents the main constraints that must be taken into account when designing time synchronization protocols for WSNs. Chapter 3 focuses on the most representative synchronization protocols for wireless sensor networks by outlining their main features, while Chapter 4 discusses three general packet-based synchronization approaches for WSNs. Chapters 5–14 propose a variety of statistical signal processing algorithms for improved estimation of clock phase offsets and assessing performance benchmarks. A series of novel results, extensions, and interesting relationships between different estimation schemes are reported. At first, the emphasis is put on developing efficient clock offset estimators for the general scenarios when the network delays are normally or exponentially distributed. However, observing the facts that the distribution of network delays in general cannot be predicted accurately, and that the estimators reported in the literature are not robust with respect to the unknown possibly time-varying distribution of network delays, we address in Chapter 14 the problem of building clock offset estimators that are robust to the distribution of network delays. Finally, Chapter 15 concludes this book with some open research problems.

This work would have not been possible without the support and encouragement of Dr. Bruce Suter from Air Force Research Laboratory (AFRL), Rome, NY, and the help of our students. Our heartfelt thanks go to Bruce, and to our families and students (Dr. Kyoung-Lae Noh, Dr. Jangsub Kim, Dr. Yik-Chung Wu, Mr. Ilkay Sari, Mr. Jaehan Lee, and Mr. Sabit Ekin). Despite our efforts to contain all sources of errors and misunderstandings, we believe that inconsistencies and errors might show up. Therefore, we are asking our readers kindly to email us their feedback to: eserpedin@gmail.com. Any feedback is welcome to improve this work. This book webpage will be maintained at http://www.ece.tamu.edu/~serpedin/.

Chapter 1

INTRODUCTION

1.1 Wireless Sensor Networks

With the help of technological advances in micro-electro-mechanical systems (MEMS) and wireless communications, low-cost, low-power, and multi-functional wireless sensing devices have been developed. When these devices are deployed over a wide geographical region, they can collect information about the environment and efficiently collaborate to process such information by forming a distributed communication network, called a *wireless sensor network* (WSN), as illustrated in Figure 1.1. A WSN is a special case of an ad-hoc wireless network, and assumes a multi-hop communication framework with no common infrastructure, where the sensors spontaneously cooperate to deliver information by forwarding packets from a source to a destination. The number of practical applications involving WSNs keeps growing rapidly, and WSNs have been regarded as providing the fundamental infrastructure for future communications due to a variety of promising potential applications: monitoring the health status of humans, animals, plants, and the environment; control and instrumentation of industrial machines and home appliances; homeland security; detection of chemical and biological threats and leaks, etc. [2], [14], [75], [100].

When designing sensor networks, there are a number of important factors to be considered such as tolerance to node failures, scalability, dynamic network topology, hardware constraints, production cost, and power consumption [2]. In general, the lifetime of a sensor network is proportional to that of a battery since the sensor nodes are usually inaccessible after deployment. Moreover, due to the space limitations and other practical constraints in sensor nodes, power is a scarce resource for practical WSNs. For these reasons, energy efficiency in general has top priority when designing WSNs out of all the above mentioned design considerations. Data

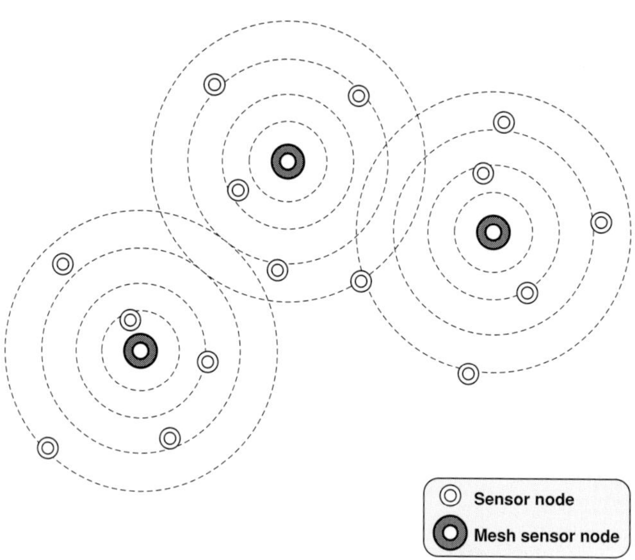

Figure 1.1: A wireless sensor network.

communication is one of the most significant operations in WSNs and requires a huge portion of the overall energy consumption. Indeed, the energy required for data communication is far greater than the energy required for data processing in a sensor node [75].

1.2 Time Synchronization

In distributed systems, maintaining the logical clocks of the computers in such a way that they are never too far apart is one of the most complex problems of computer engineering. Whether disciplining computer clocks with the devices synchronized to a global positioning system (GPS) satellite or a network time protocol (NTP) [62] time server over the Internet, it is possible to equip some primary time servers to synchronize a much larger number of secondary servers and clients connected through a common infrastructure. In order to do this, a distributed network clock synchronization protocol is required through which a server clock can be read, the readings to other clients can be transmitted, and each client clock can be adjusted as required. In such a distributed synchronization approach, the participating devices exchange timing information with their chosen reference at regular intervals and adjust their logical clocks accordingly.

A computer clock in general has two components, namely a frequency source and a means of accumulating timing events (consisting of a clock interrupt mechanism and a counter implemented in software). The implementation of the computer clock in the operating system and the programming interface differ between operating systems and hardware platforms. However, the basic sources of timing errors are an uncompensated quartz crystal oscillator and the clock interrupts it generates. Theoretically, two clocks would remain synchronized if their offsets were set equal and their frequency sources run at the same rate. However, in practice clocks are set with limited precision and the frequency sources run at slightly different rates. In addition, the frequency of a crystal oscillator varies due to the initial manufacturing tolerance, aging, temperature, pressure, and other factors. Because of these inherent instabilities, distributed clocks must regularly be synchronized.

1.3 Importance of Time Synchronization

Time synchronization is a procedure for providing a common notion of time across a distributed system. It is crucial for WSNs when performing a number of fundamental operations, such as:

- *Data fusion* Data merging is a major operation in all distributed networks for processing and integrating the collected data in a meaningful way, and it requires some or all nodes in the network to share a common timescale.
- *Power management* Energy efficiency is a key factor when designing WSNs since sensors are usually left unattended without maintenance and battery replacement for their lifetimes after deployment. Most energy-saving operations strongly depend on time synchronization. For instance, duty cycling (sleep and wake-up modes control) helps the nodes to save huge energy resources by spending minimal power during the sleep mode. Thus, network-wide synchronization is essential for efficient duty cycling and its performance is proportional to the synchronization accuracy.
- *Transmission scheduling* Many scheduling protocols require time synchronization. For example, the time division multiple access (TDMA) scheme, one of the most popular communications schemes for distributed networks, is only applicable to a synchronized network.
- *Miscellaneous* Many localization, security, and tracking protocols also demand the nodes to timestamp their messages and sensing events.

Therefore, time synchronization is one of the most important research challenges in the design of energy-efficient WSNs.

1.4 History of Clock Synchronization

Many time synchronization protocols for maintaining synchronization of physical clocks over computer networks have been designed over the past few decades. The pioneering work in this area was the remote clock reading method proposed by Cristian, which handles unbounded message delays between processes [16]. In this algorithm, a time request is sent to the remote process and after receiving the response, the host process calculates the round-trip time as the difference between the transmission and reception times. The main feature of this algorithm relies on the fact that performing a large number of request/response experiments will make it more likely that at least one trial will not encounter random delays. Logically, the chosen trial is the one with the least round-trip delay.

The standard for time synchronization in the Internet, the network time protocol (NTP), was developed by Mills [62] (also see [61]). NTP is a protocol for synchronizing the clocks of computer systems over packet-switched, variable-latency data networks. It is a layered client–server architecture based on user datagram protocol (UDP) message passing, which synchronizes computer clocks in a hierarchical way. The main attractive features of NTP are its scalability, robustness to failure, self-configuration in large multi-hop networks, and ubiquitous deployment. NTP's sender–receiver synchronization architecture is now widely accepted in designing time synchronization algorithms and relies on the same two-way timing message exchange mechanism targeted in most of this book.

The time transmission protocol (TTP), which is used by a node to communicate the time of its clock to a target node, was introduced in [7]. The target node estimates the time in the source node by using the message timestamps and message delay statistics without any feedback response or pairwise synchronization. The exploitation of the inherent properties of the broadcast media, where a node sends timing beacons to all the nodes within a single broadcast domain, was first introduced in the CesiumSpray system [97]. Later, one of the most popular clock synchronization protocols in wireless sensor networks, the reference broadcast synchronization (RBS) [22], was proposed based on the same idea, but allowing an extension of the protocol to multiple domains. The RBS protocol will be discussed in detail in Chapter 3. For ad-hoc communication networks, the time synchronization protocol [79] is one of the pioneering contributions. This protocol is based on generating timestamps to record the time at which an event of interest occurred. The timestamps are updated by each node using its local clock and the time transformation method, in which the final timestamp is expressed in terms of an interval with a lower bound and an upper bound. Clock synchronization in WSNs is an entirely new area of research due to the fact that the above general synchronization protocols present a number of challenges when applied to WSNs. This is because of the unique nature of sensor networks: limited power resources, wireless channel conditions, and dynamic topology caused by node mobility and failure. Therefore, different types of synchronization schemes have to be designed explicitly for WSN applications to cope with these challenges. References [23], [24], [80], [81], [82], [84], [88], [89], [91] are excellent surveys of research in this area.

The main protocols put forward for clock synchronization to cope with the above mentioned requirements in WSNs are described next. RBS [22] is based on the post-facto receiver–receiver synchronization approach. In RBS, a reference broadcast message is sent by a node to two or more neighboring nodes which record their own local clocks at the reception of broadcasted message. After collecting a few readings, the nodes exchange their observations and a linear regression approach is used to estimate their relative clock offset and skew. The timing synch protocol for sensor networks (TPSN) [29] is a conventional sender–receiver protocol which assumes two operational stages: the level discovery phase followed by the synchronization phase. During the level discovery phase, the WSN is organized in the form of a spanning tree, and global synchronization is achieved by enabling each node to be synchronized with its parent (the node located in the adjacent upper level) by means of a two-way message exchange mechanism through adjusting only its clock offset. Because TPSN and NTP share the same signaling mechanism, one might interpret TPSN as an extension of NTP to WSNs.

The timing synchronization protocol for high latency acoustic networks (TSHL) [93] combines both of these approaches, namely the receiver–receiver (RBS) and sender–receiver (TPSN), in two stages. The first stage in TSHL is similar to RBS while the second stage is similar to TPSN. TSHL is particularly suitable for networks involving large message delays, e.g., underwater acoustic networks. The flooding time synchronization protocol (FTSP) [58] also combines the two approaches in the sense that the beacon node sends its timestamps within the reference broadcast messages. The delay measurement time synchronization (DMTS) protocol for WSNs [74] is an energy-efficient protocol which avoids estimating the round-trip time unlike NTP or TPSN, synchronizes the sender and multiple receivers at the same time, and requires a smaller number of message transfers than RBS. Among other non-probabilistic approaches, the tiny–mini-synchronization protocols [86] provide good synchronization accuracy in WSNs while using a deterministic protocol with minimal computational and storage complexity.

The Time Diffusion Protocol (TDP) proposed in [90] achieves a network-wide time equilibrium by using an iterative, weighted averaging technique based on a diffusion of messages involving all the nodes in the synchronization process. Also, the asynchronous diffusion protocol (ADP) [53] uses a diffusion strategy similar to TDP, but the network nodes execute the protocol and correct their clocks asynchronously with respect to each other.

Suggesting a probabilistic approach to tackle the clock synchronization problem in WSNs, PalChaudhuri *et al.* [69] extended the results from RBS by providing probabilistic bounds on the accuracy of clock synchronization. Valuable research was carried out by Abdel-Ghaffar [1] who adopted the probabilistic approach. He presented a detailed analysis of clock offset estimation assuming a symmetric exponential model for network delays. It was implicitly argued in [1] that for known fixed delay and exponential delay parameters, the maximum likelihood estimator of the clock offset does not exist because the likelihood function does not possess a unique maximum with respect to the clock offset. However, it was proved by Jeske in [41] that for an unknown fixed delay, irrespective of the exponential delay parameter being known or unknown, the MLE of the clock offset does exist and coincides with a previously proposed estimator by Paxson in [71] and [72].

The research presented in this book advances the results for the clock synchronization problem for WSNs by adopting a probabilistic approach and applying techniques from statistical signal processing and estimation theory. The next section lists the major topics in the book. Some very interesting results for this problem were derived in [27], which showed that for the synchronization problem consisting of estimating all the unknown parameters, i.e., skews and offsets of all the clocks as well as the delays of all the communication links, even in the presence of noiseless communication of messages, the estimation of all the unknown parameters is impossible.

1.5 OUTLINE

The topics covered in this book are summarized as follows.

In Chapter 2, the general clock model for time synchronization is first introduced and analyzed. The most important features that have to be considered when designing time synchronization protocols for WSNs are presented. In addition, the different delay components that are present in the timing message delivery are categorized. In Chapter 3 we categorize and survey the existing time synchronization protocols for WSNs, focusing mainly on the signal processing aspects. In addition, in Chapter 3 we describe the importance and effectiveness of adaptive time synchronization schemes, and introduce some important adaptive synchronization protocols as well. In Chapter 4 three general and fundamentally different time synchronization approaches are presented, based on packet synchronization, namely, sender–receiver, receiver–receiver, and receiver-only synchronization. These basic approaches are analyzed and compared to illustrate the common and different characteristics in clock synchronization of WSNs.

In Chapter 5, the best linear unbiased estimates using order statistics (BLUE-OS) of the clock offset between two nodes for a sender–receiver timing exchange paradigm are derived assuming both symmetric and asymmetric exponential network delays. The Rao–Blackwell–Lehmann–Scheffé theorem (see e.g., [44], [50]) is then exploited to obtain the minimum variance unbiased estimate (MVUE) for the clock offset and is shown to coincide with the BLUE-OS. In addition, it is found that the MVUE of the clock offset in the presence of symmetric network delays also coincides with the maximum likelihood estimate (MLE). Finally, in the presence of asymmetric network delays, although the MLE is biased, it is shown to achieve lower mean square error (MSE) than the MVUE in the region around the point where the bidirectional network link delays are symmetric and hence its merit as the most versatile estimator is fairly justified. Chapter 6 is focused on analyzing and deriving the MLEs and the corresponding Cramer–Rao lower bounds (CRLBs) for the conventional clock offset model in a sender–receiver timing exchange assuming a Gaussian model for the noise. Next, the joint MLE and corresponding CRLB using a more realistic linear clock offset and skew model assuming Gaussian random delays are also obtained. The MLEs for the clock-offset-only case, and joint estimation of the clock offset and skew under the exponential delay assumption are then derived and the corresponding algorithms for finding these estimates are also presented in detail. Since the MLEs are not robust and are computationally complex, simpler algorithms, even those which exhibit some performance degradation, are more suited for the low-power constraints of WSNs. Therefore, in Chapter 7 we discuss two simplified schemes to estimate both the clock offset and skew and that require negligible computations. In the first scheme the first and the last sample of the observations are utilized and the estimators are derived under both the Gaussian and exponential delay models, while in the second scheme a line between two points is fitted at the minimum distance apart regardless of the actual delay distribution involved. Simulation results are also presented and compared with the MLE results.

Chapter 8 discusses a novel time synchronization scheme referred to as the pairwise broadcast synchronization (PBS) protocol [66], [68], which efficiently combines the sender–receiver synchronization and receiver-only synchronization approaches to achieve network-wide synchronization with a significantly reduced number of synchronization messages, i.e., with less energy consumption. The extension of PBS to general multi-cluster sensor networks is also studied in this chapter. Extending the idea of having the inactive nodes in a WSN overhear the two-way timing message communication between two active (master and slave) nodes, in Chapter 9 we derive the MLE for the clock offsets of the inactive nodes located within the communication range of the active nodes by assuming an exponential link delay modeling, hence synchronizing with the reference node at essentially zero cost. A vital implication of this work is that the performance of the sender–receiver protocols, whose main disadvantage has always been categorized as the high communication overhead in WSN scenarios due to their point-to-point rather than broadcast nature, can be compared with that of receiver–receiver protocols on equal grounds. In addition, the CRLBs for both the active and inactive nodes are also obtained as a performance benchmark. In Chapter 10 we advance the results from Chapter 9 in two domains. First, the BLUE-OS is derived by applying general least-squares theory to an ordered sample. The MVUE is also obtained by the application of Rao–Blackwell–Lehmann–Scheffé theorem, which is shown to coincide with the BLUE-OS. In addition, since the MSE is usually selected as the performance criterion in estimation theory, a minimum mean square error (MMSE) estimator with expected loss independent of the clock offset and fixed delay is also derived and shown to outperform the MVUE. Second, the results presented are generalized by addressing the problem for both symmetric and asymmetric exponential delays, since in practice the message exchange involves asymmetry in general for each direction.

In Chapter 11, an energy-efficient adaptive multi-hop timing synchronization (AMTS) scheme with the goal of achieving a long-term network-wide synchronization with minimal energy consumption is described. Since many applications require long-term synchronization among the nodes, and experiments have shown that a quadratic model of clock variations can better capture the dynamics of the actual clock model involved, the MLE for all the clock parameters in a two-way timing exchange model with exponential delays is derived in Chapter 12.

Shifting our focus towards a receiver–receiver protocol, in Chapter 13 we obtain the joint MLE for the clock offset and skew under the exponential noise model. The Gibbs sampler is also proposed for joint clock offset and skew estimation and shown to provide superior performance than the joint maximum likelihood estimator (JMLE). Lower and upper bounds for the MSEs of the JML-estimator and Gibbs sampler are introduced in terms of the MSE of the MVUE and the conventional BLUE, respectively.

In Chapter 14 we address the problem of robust estimation of clock offset parameters for synchronization protocols that rely on the two-way message exchange mechanism and operate in the presence of unknown distributions for the network delays. Within the framework of sequential Bayesian Monte-Carlo approaches, several families of clock offset estimators are proposed that are shown to be robust to the distribution of the network delays and to exhibit superior performance relative to the classical linear regression and MLE approaches.

Finally, in Chapter 15 we summarize the results presented in the book with concluding remarks and also formulate some possible future research directions. The fact that the results presented in this book are applicable to a wide range of time transfer problems either directly or through some minor extensions is emphasized.

Chapter 2

SIGNAL MODELS FOR TIME SYNCHRONIZATION

2.1 Definition of Clock

Every individual sensor in a network has its own clock. The counter in a sensor is increased in accordance with the zero-crossings or the edges of the periodic output signal of the local oscillator. When the counter reaches a certain threshold value, an interrupt is created and delivered to the memory. The frequency of the oscillator and the threshold value determine the resolution of the clock. Ideally, the clock of a sensor node should be configured such that $C(t) = t$, where t stands for the ideal or reference time. However, due to the imperfections of the clock oscillator, the clock function of the ith node is modeled as

$$C_i(t) = \phi + \omega t + \epsilon, \tag{2.1}$$

where the parameters ϕ and ω are called the clock offset (phase difference) and clock skew (frequency difference), respectively, and ϵ stands for random noise.

Assuming the effect of random noise ϵ is negligible, from (2.1), the clock relationship between two nodes, say node 1 and node 2, can be represented by

$$C_1(t) = \phi^{(12)} + \omega^{(12)} C_2(t),$$

where $\phi^{(12)}$ and $\omega^{(12)}$ are the relative clock offset and skew between node 1 and node 2, respectively. Thus, $\phi^{(12)} = 0$ and $\omega^{(12)} = 1$ when the two clocks are perfectly synchronized. Suppose there are L nodes in the network, then the global network-wide synchronization is achieved when $C_i(t) = C_j(t)$ for all $i, j = 1, \ldots, L$.

Time synchronization in WSNs is a complicated problem due to the following reasons. First, every single oscillator has its unique clock parameters regardless of its type. For instance, according to the data-sheet of a typical crystal-quartz oscillator commonly used in sensor networks, the frequency of a clock varies up to 40 parts per million (ppm), which means clocks of different nodes can lose as much as 40 ms in a second. In other words, every single oscillator might assume a different skew parameter ranging from -20 to 20 ppm.

Notice that, in general, the clock skew ω is a time-dependent random variable (RV) and there are two concepts that are often used in clock terminology regarding the nature of the time-dependent randomness present in clock parameters. These concepts are referred to as *short-term* and *long-term* stabilities. Short-term instability is primarily due to environmental factors, e.g., temperature variations, shock and supply voltage, whereas long-term instability results from more subtle effects, like oscillator aging [22]. For the oscillators currently used in sensor networks, all these parameters are almost constant for short time intervals [98]. Moreover, the total power of the noise process is too small to be effective in short time-spans [38]. Therefore, the parameters of a clock are assumed to be constant for the time period of interest.

As far as the long-term stability is concerned, the clock parameters are subject to changes due to environmental or other external effects such as temperature, atmospheric pressure, voltage changes, and hardware aging [98]. Hence, in general, the relative clock offset keeps changing with time, which means that the network has to perform periodic time resynchronization to adjust the clock parameters.

2.2 Design Considerations

Time synchronization for conventional wired networks has been thoroughly studied and a plethora of synchronization protocols have been proposed as surveyed in [2]. However, for WSNs, there are a number of unique and important factors to be considered when designing time synchronization protocols as listed next.

- *Energy consumption* Energy consumption is a very important consideration in WSNs due to their limited and generally non-rechargeable power resources [25]. Hence, WSNs should be designed to maintain minimal energy expenditure in each sensor node. Various types of power control procedures, such as sleep/wake-up modes and dynamic routing controls, are commonly considered in this regard. Time synchronization is one of the critical components contributing to energy consumption due to the high energy-consuming radio transmissions needed to achieve clock synchronization. One study suggests that the energy consumption required for time synchronization of a node is approximately 17% of the total energy spent by a node [32]. Pottie and Kaiser showed in [75] that the radio frequency (RF) energy required to transmit 1 bit over 100 meters (i.e., 3 joules) is roughly equivalent to the energy required to execute 3 million instructions. Therefore, developing efficient synchronization algorithms represents an ideal mechanism for trading computational energy for reduced RF communication energy. In the sequel, energy efficiency is the main concern in designing time synchronization protocols.

- *Latency* Latency in message delivery is a fundamental factor when designing communications networks. For networks that rely on multi-hop transmissions like WSNs, latency is even more critical because the uncertainty in message delivery significantly increases as the number of hops increases. Moreover, the effects of channel variations, mobility, and the ad-hoc nature of WSNs make the synchronization problem more complex. Efficient localization and time synchronization protocols are necessary to reduce the latency error and jitter.

- *Security and reliability* Network security has received much attention as the networks have become more accessible and vulnerable due to the development of sophisticated spying techniques and devices. In addition, unlike for wired networks, far more frequent message losses occur in wireless networks because of the time-varying nature of wireless channels. Therefore, a mechanism to cope with message losses and malicious attacks in time synchronization is necessary for WSNs. Preliminary results in this regard have been proposed and discussed in [12], [20], [65].

- *Network topology changes* The performance of a time synchronization protocol is closely related to the network topology, i.e., it varies with respect to the density and distribution of sensors in the network. Therefore, any shift in the location or scale of sensors incurs a network topology change, which requires in its turn a new self-configuration. Mobility of the sensors and battery timeouts are the main reasons for this change. Hence, for dynamic sensor networks, time synchronization protocols should be able to adapt well to frequent network topology changes.

- *Scalability* Scalability is another important factor in the design of synchronization protocols. The computational and implementation complexity of synchronization algorithms becomes a critical problem as the number of sensors is very large. Therefore, it is highly desirable for the clock synchronization algorithms to be scalable with an increase in the number of network nodes. In addition, many other crucial MAC operations, such as multi-hop routing and network configuration, greatly depend on the network scalability.

2.3 Delay Components in Timing Message Delivery

The main roles of time synchronization in a distributed network are to ensure a common timescale for all the network nodes, and to provide the right temporal coordination among all the nodes engaged in a collaborative and distributed interaction with the physical environment. Timing mismatch arises mainly from different setup times of nodes and time variations introduced by local oscillators running at different frequencies. Environmental variations, such as temperature and aging, also cause local clock oscillators to run unpredictably. All these uncertainties cause the local clocks of different nodes to drift apart over the course of a time interval.

Assume that two nodes need to be synchronized with respect to one another. One of the nodes sends its current time to the other node; then if there is absolutely no delay in the message delivery, that neighboring node will immediately know the difference between its clock and its neighbor's clock. Unfortunately, in a real wireless network, various delays affect the message delivery, making time synchronization much more difficult than it appears at first sight. In general, a series of timing message transmissions is required to estimate the relative time offsets between nodes. In some ways, time synchronization in wireless sensor networks can be regarded as a process of removing the non-deterministic delays during timing message transmission over wireless channels.

There are a number of non-deterministic delays while transferring messages between nodes. Kopetz and Ochsenreiter were the first to analyze the structure of message delays and characterized the delay components according to the process of message delivery [45]. The delay components in message delivery can be categorized as follows.

1. Send time: the time spent in building the message at the application layer including other delays introduced by the operating system when processing the send request. The send time is non-deterministic and can be up to hundreds of milliseconds, depending on the workload of the system.

2. Access time: the waiting time for accessing the channel after reaching the MAC layer. This is the most significant factor and is highly variable depending on the specific MAC protocol used. The access time is non-deterministic and varies from milliseconds up to seconds, depending on the current network traffic.

3. Transmission time: the time to transmit a message at the physical layer. This delay can be estimated from the length of a message and the speed of radio in the medium and is of the order of tens of milliseconds.

4. Propagation time: the actual time taken to transmit a message from the sender to the receiver through the wireless channel. The propagation time is deterministic and, in general, is less than 1 microsecond, which is almost negligible compared with the other delay components.

5. Reception time: the time required to receive a message at the physical layer, which is the same as the transmission time. In some cases, this delay has been categorized as a part of the receive time.

6. Receive time: the time required to construct and send a received message to the application layer at the receiver. It can be viewed as the corresponding component at the receiver side of the send time at the transmitter side, and can be time-varying due to the variable delays introduced by the operating system.

Note that the time delay in message transmission also depends on other factors, such as the hardware platform, the error correction code, and the modulation scheme. The estimated time delay discussed above in each component is based on the MICA platform, [34], [60]. A more detailed analysis can be found in [58].

Chapter 3

TIME SYNCHRONIZATION PROTOCOLS

Ideally, a time synchronization protocol should be able to work optimally in terms of all the design requirements imposed on time synchronization in WSNs, which include energy efficiency, scalability, precision, security, reliability, and robustness to network dynamics. However, the complex nature of WSNs makes it very difficult to optimize the protocol with respect to all these requirements simultaneously. Due to the tradeoffs in satisfying these requirements, each protocol is designed to put distinct emphases on different requirements.

Assuming various criteria, time synchronization protocols can be categorized into different classes:

- *Master–slave vs. peer-to-peer*
  - *master–slave*: first a tree-like network hierarchy is arranged, and upon the completion of this arrangement only the connected nodes in the hierarchy synchronize with each other.
  - *peer-to-peer*: any pair of nodes in the network can synchronize with each other.
- *Clock correcting vs. untethered clock*
  - *clock correcting*: the clock function in memory is modified after each run of the time synchronization process.
  - *untethered clock*: every node maintains its own clock as it is, and keeps a time-translation table relating its clock to other nodes' clocks; thus, instead of updating its clock constantly, each node translates the time information in the data packets coming from other nodes to its own clock by using the time-translation table.

- *Synchronization approach*
  - *sender–receiver*: one of two nodes, which are synchronizing with each other, sends a timestamp message while the other one receives it.
  - *receiver–receiver*: a reference node transmits synchronization signals and two synchronizing nodes receive these signals and record the reception times (timestamps).
  - *receiver-only*: a group of nodes can be simultaneously synchronized by listening to the message exchanges of a pair of nodes.
- *Pairwise synchronization vs. network-wide synchronization*
  - *pairwise synchronization*: the protocols are primarily designed to synchronize two nodes, although they usually can be extended to handle synchronization of a group of nodes.
  - *network-wide synchronization*: the protocols are primarily designed to synchronize a large number of nodes in the network.

Additional classifications can be found in [91]. In the following, we will summarize the existing time synchronization protocols based on the last category.

3.1 PAIRWISE SYNCHRONIZATION

3.1.1 TIMING-SYNC PROTOCOL FOR SENSOR NETWORKS (TPSN)

TPSN [29] assumes a sender–receiver synchronization (SRS) approach that exploits the two-way message exchange mechanism, depicted graphically in Figure 3.1, to achieve the synchronization between two nodes. The SRS approach will be described in detail in Chapter 4, where comparisons and relationships with other signaling mechanisms to achieve time synchronization will be also presented. Figure 3.1 presents a number of N two-way message exchanges between two nodes, node A and node P, which are assumed to exhibit the clock phase offset difference $\phi^{(AP)}$. As illustrated in Figure 3.1, the timestamps $T_{1,i}^{(A)}$, $T_{4,i}^{(A)}$, $T_{2,i}^{(P)}$ and $T_{3,i}^{(P)}$ are made during the ith message exchange measured by the local clocks of node A and node P, respectively. Furthermore, from Figure 3.1, the time stamps, $T_{2,i}^{(P)}$ and $T_{4,i}^{(A)}$, can be expressed as

$$T_{2,i}^{(P)} = T_{1,i}^{(A)} + \phi^{(AP)} + d^{(AP)} + X_i^{(AP)}, \tag{3.1}$$

$$T_{4,i}^{(A)} = T_{3,i}^{(P)} + \phi^{(PA)} + d^{(PA)} + X_i^{(PA)}, \tag{3.2}$$

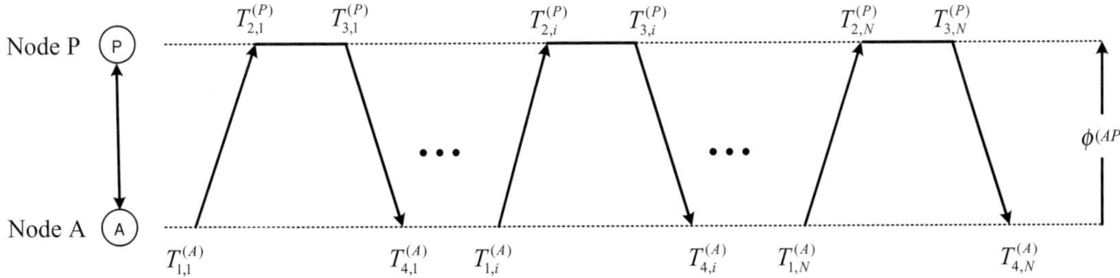

Figure 3.1: Clock synchronization model of SRS.

where $d^{(AP)}$ and $X_i^{(AP)}$ are the fixed and random portions of timing delays in the message transmissions between nodes A and P, respectively. Also the time differences are defined as $U_i \triangleq T_{2,i}^{(P)} - T_{1,i}^{(A)}$ and $V_i \triangleq T_{4,i}^{(A)} - T_{3,i}^{(P)}$.

Assuming that the two nodes A and P exchange only one round of message exchanges ($N = 1$), and without statistical model for the variable delay components $X_1^{(AP)}$ and $X_1^{(PA)}$ in (3.1) and (3.2), a simple estimate for $\phi^{(AP)}$ is proposed in [29] as

$$\hat{\phi}^{(AP)} = \frac{U_1 - V_1}{2}, \tag{3.3}$$

where $U_1 \triangleq T_{2,1}^{(P)} - T_{1,1}^{(A)}$ and $V_1 \triangleq T_{4,1}^{(A)} - T_{3,1}^{(P)}$. Notice that the original form of TPSN does not estimate the clock skew. Therefore, frequent application of TPSN is needed to keep the clock offset between the two nodes below a certain limit. Assume the clock offset $\phi^{(AP)}$ is constant for N rounds of message exchanges. If $X_i^{(AP)}$ and $X_i^{(PA)}$ are exponentially distributed with the same unknown mean λ and assuming that $d \triangleq d^{(AP)} = d^{(PA)}$ is unknown, it is proved in [41] that the maximum likelihood (ML) estimator of $\phi^{(AP)}$ is given by

$$\hat{\phi}^{(AP)} = \frac{\min_{1 \le i \le N} U_i - \min_{1 \le i \le N} V_i}{2}. \tag{3.4}$$

On the other hand, with $X_i^{(AP)}$ and $X_i^{(PA)}$ being modeled as independent and normally distributed random variables (RVs) with the same mean μ and variance $\sigma^2/2$, the ML estimate for $\phi^{(AP)}$ takes the expression (derived in Chapter 4)

$$\hat{\phi}^{(AP)} = \frac{\frac{1}{N}\sum_{i=1}^{N} U_i - \frac{1}{N}\sum_{i=1}^{N} V_i}{2}. \tag{3.5}$$

From (3.3), (3.4), and (3.5), it is clear that if only one round of message exchange is performed, the TPSN time offset estimate shown in (3.3) is the ML estimator under both exponential and Gaussian delay models. However, in general, the two estimates

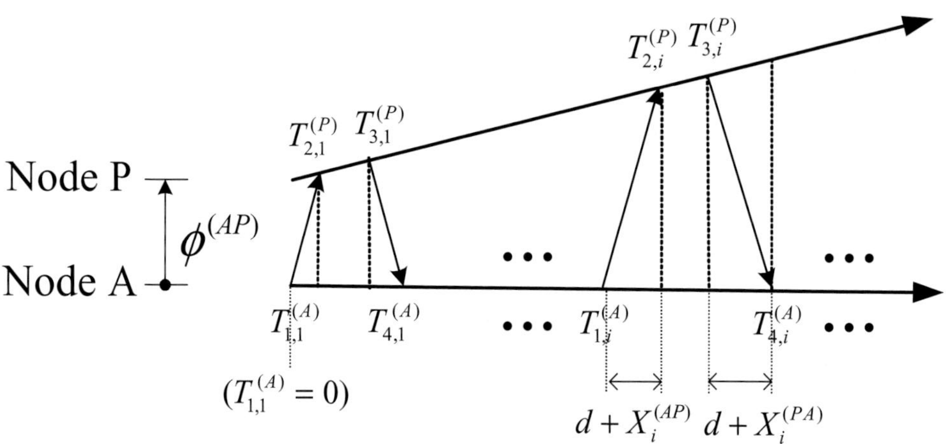

Figure 3.2: Linear clock skew model for message exchanges.

shown in (3.4) and (3.5) are different. Chapter 4 describes in detail all these results for the sender–receiver synchronization approach.

In Chapter 6, we will propose a practical joint clock offset and skew correction scheme to guarantee the long-term stability of synchronization for TPSN. Moreover, the joint offset and skew ML estimators for TPSN under both Gaussian and exponential delay models will be also derived in Chapter 6.

3.1.2 Tiny-Sync and Mini-Sync

Tiny-sync and mini-sync [86] are two lightweight clock synchronization protocols that also rely on the two-way message exchange mechanism. Suppose that node A and node P exchange timing messages as in Figure 3.2. This figure shows the effect of clock offset (ϕ) and skew (ω) on timing message exchanges between the two nodes. Without loss of generality, the reference time $T_{1,1}^{(A)}$ is set to be zero. Here, the timestamp at node P in the ith uplink message $T_{2,i}^{(P)}$ is given by

$$T_{2,i}^{(P)} = \omega^{(AP)}(T_{1,i}^{(A)} + d + X_i^{(AP)}) + \phi^{(AP)}, \tag{3.6}$$

where the term $\omega^{(AP)}(T_{1,i}^{(A)} + d + X_i^{(AP)})$ is due to the effect of clock skew. Similarly, the timestamp at node P in the ith downlink message $T_{3,i}^{(P)}$ takes the expression

$$T_{3,i}^{(P)} = \omega^{(AP)}(T_{4,i}^{(A)} - d - X_i^{(PA)}) + \phi^{(AP)}, \tag{3.7}$$

where the term $\omega^{(AP)}(T_{4,i}^{(A)} - d - X_i^{(PA)})$ is again due to the effect of clock skew.

Assume for the moment that node P replies to node A immediately after receiving the message, i.e., $T_{2,i}^{(P)} = T_{3,i}^{(P)}$. From (3.6) and (3.7), we infer that

$$\frac{T_{2,i}^{(P)} - \phi^{(AP)}}{\omega^{(AP)}} = T_{1,i}^{(A)} + d + X_i^{(AP)},$$
$$\frac{T_{2,i}^{(P)} - \phi^{(AP)}}{\omega^{(AP)}} = T_{4,i}^{(A)} - d - X_i^{(PA)}.$$

Since d, $X_i^{(AP)}$, and $X_i^{(PA)}$ are all non-negative, defining $\omega' \triangleq 1/\omega^{(AP)}$ and $\phi' \triangleq \phi^{(AP)}/\omega^{(AP)}$, we obtain

$$T_{1,i}^{(A)} \le \omega' T_{2,i}^{(P)} + \phi' \le T_{4,i}^{(A)}. \tag{3.8}$$

The 3-tuple of timestamps ($T_{1,i}^{(A)}, T_{2,i}^{(P)}$, and $T_{3,i}^{(A)}$) is called a data point. With N message exchanges, the goal is to find ϕ' and ω' such that they satisfy (3.8) for $i = 1, \ldots, N$. In general, this is a linear programming problem and there are an infinite number of solutions for this problem [51]. Although more timestamps would generate tighter bounds on ϕ' and ω', unfortunately, at the same time, the computational and storage requirements of the linear programming approach also increases. Thus, such an approach does not appear suitable to be implemented in wireless sensor nodes, which have strictly limited memory and computing resources.

Tiny-sync and mini-sync tackle the problem by finding the best-fit line that lies between the bound sets defined by the data points. Based on the observation that not all data points are useful, Tiny-sync preserves only four constraints (the ones that yield the best bounds on the estimate) out of all data points. This results in a very efficient algorithm. However, it has been shown by a counterexample [86] that this scheme does not always produce the optimal solution since some data points are considered useless and are discarded at a certain time, a step which might actually provide a better bound if it were properly considered with another data point that is yet to come.

Mini-sync is an improved version of tiny-sync in the sense that it finds the optimal solution with increased complexity (but still with less complexity than the linear programming approach). Mini-sync basically uses an additional criterion to determine whether the data point can be safely discarded.

3.1.3 REFERENCE BROADCAST SYNCHRONIZATION (RBS)

The basic signaling mechanism used in RBS is depicted graphically in Figure 3.3. RBS [22] is based on the receiver–receiver synchronization (RRS) approach that will be discussed in more detail in Chapter 4. In Figure 3.3, node P represents the parent or reference node, which initiates the synchronization protocol by transmitting

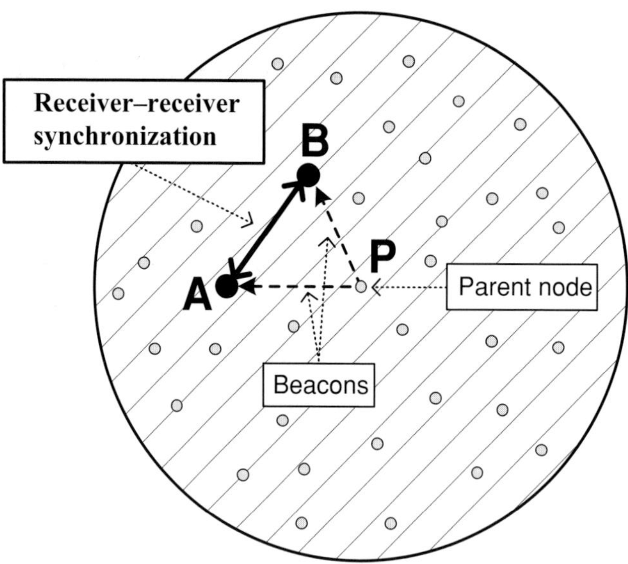

Figure 3.3: RBS signaling mechanism.

beacons to node A and node B, which are assumed to lie within the communication range of node P. Let the timestamps recorded at node A and node B for receiving the ith common packet be denoted as $T_{2,i}^{(A)}$ and $T_{2,i}^{(B)}$, respectively. The estimate of the clock offset between node A and node B is proposed in [22] as

$$\hat{\phi}^{(BA)} = \frac{1}{N} \sum_{i=1}^{N} \left[T_{2,i}^{(A)} - T_{2,i}^{(B)} \right], \tag{3.9}$$

where N stands for the total number of common packets received by node A and node B. In Chapter 4 we show that the above estimator is actually the ML estimator for the clock offset, assuming the random portions of the delays in message deliveries are Gaussian distributed RVs, and there is no clock skew. When there is a clock skew between node A and node B, least-squares linear regression is proposed in [22] to estimate the clock skew.

The main advantage of RBS is that by comparing the timestamps of a common packet at two different nodes, it removes the largest sources of non-deterministic error (send time and access time) from the transmission path. Thus, RBS provides a high degree of synchronization accuracy. Note also that RBS can be applied to commodity hardware and existing software in sensor networks as it does not need access to the low levels of the operating system.

When a sensor node observes and synchronizes to a broadcast clock, Sadler [83] derived the ML estimator for clock offset and skew with the broadcast message delay being modeled as uniformly distributed RVs. He showed that the ML estimate in this case is generally not unique. Furthermore, the support of likelihood function is not convex which removes the possibility of taking the mean of all equally likely solutions. This motivated Sadler to consider the linear estimator for the clock offset and skew. For the same case, in Chapter 13 we will infer the joint ML clock offset and skew estimator under the assumption that the broadcast message delays are modeled as exponentially distributed RVs. It will be shown in Chapter 13 that a unique joint ML clock offset and skew estimate exists under certain conditions, as opposed to the case of uniformly distributed delay. Furthermore, the Gibbs sampler will be introduced in Chapter 13 to enhance the performance of the joint ML estimator further.

3.1.4 Flooding Time Synchronization Protocol (FTSP)

In [58], it is argued that if one can timestamp the message at the MAC layer, this immediately eliminates three sources of delay uncertainties: transmit, access, and receive times. In this case, the main delivery delay comes from the transmission and reception times at the radio chips (see Chapter 2). These delays can be further decomposed into: (1) the interrupt handling time, which is the delay between the radio chip raising and the microcontroller responding to an interrupt; (2) the encoding time, which is the time it takes for the radio chip to encode and transform the message into a radio wave; (3) the decoding time, which is the time it takes for the radio chip at the receiver to transform the radio wave back into binary data; and (4) the byte alignment time, which is the delay at the receiver due to synchronizing with the byte boundary at the physical layer.

FTSP [58] uses a single broadcasted message to establish synchronization points between the sender and the receivers, while eliminating the jitter of interrupt handling and encoding/decoding times by utilizing multiple MAC layer timestamps on both the sender and the receiver side. Furthermore, the skew of the clock between the sender and the receiver is estimated using multiple messages and linear regression. Hence, the major difference between FTSP and RBS lies in the fact that the beacons are not timestamped at the master node in RBS protocol, which results in a large communication overhead.

3.2 Network-Wide Synchronization

Up to this point, we have only described time synchronization between two neighboring sensor nodes. In this section, we will discuss the protocols for network-wide (or global) synchronization.

3.2.1 EXTENSION OF TPSN

In order to establish a global timescale for all the nodes in the sensor field based on TPSN, Ganeriwal *et al.* [29] propose creating first a hierarchical structure (spanning tree) in the network (a stage referred to as the level discovery phase) before pairwise synchronization is performed between adjacent levels (this latter stage is referred to as the synchronization phase). The level discovery phase consists of the following steps: (1) select a root node using an appropriate leader election algorithm and assign a 0-level to the root node; (2) the root node broadcasts a level discovery packet (LDP) containing the identity and the level of the packet; (3) every node who receives an LDP assigns it a level that is one level higher than that of the received packet and sends a new level discovery packet attaching its own level; once assigned a level, a node neglects future packets requesting level discovery to avoid flooding congestion; (4) repeat step (3) until every node in the network is successfully assigned a level.

After the spanning tree is formed, the root node initiates the synchronization phase by synchronizing all the nodes in level 1. Next, the nodes in level 1 synchronize with the nodes in level 2, and so on, until all the nodes have been synchronized. Notice that the synchronization error of a node with respect to the root node is a non-decreasing function of the hop distance because the random signal errors over each hop are additive. To limit the synchronization errors, a number of different searching algorithms have been considered in the construction of the spanning tree. For instance, Van Greunen and Rabaey suggested some preliminary ideas on constructing spanning trees with low depth in order to improve the accuracy of synchronization [32].

3.2.2 LIGHTWEIGHT TIME SYNCHRONIZATION (LTS)

Also based on the two-way message exchange mechanism, in LTS [32] two network-wide synchronization protocols are proposed. The first one is called centralized multi-hop LTS, and is basically the same protocol as the extension of TPSN discussed above. The other one is called distributed multi-hop LTS. This distributed LTS algorithm moves the resynchronization from the root node to the nodes that need resynchronization. When node A determines that it needs to be resynchronized, it sends a resynchronization request to the root node. In order for node A to re-synchronize, all nodes along the routing path from the root node to node A are synchronized in a pairwise fashion. When the clock skews are bounded, LTS provides an alternative approach for synchronization that exhibits low complexity and high efficiency.

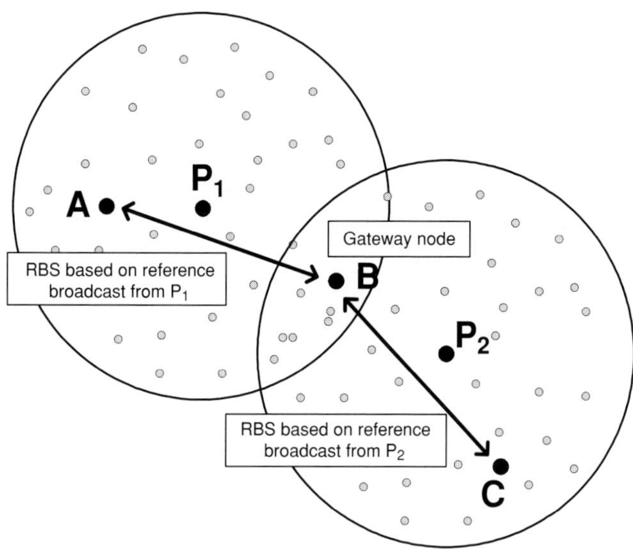

Figure 3.4: Extension of RBS to multi-hop.

3.2.3 Extension of RBS

The RBS protocol discussed in the previous subsection can only synchronize a set of nodes that lie within a single broadcast domain. In order to extend RBS to achieve global synchronization of a large-scale sensor network, Elson *et al.* [22] propose using *gateway* nodes to convert timestamps from one neighborhood's timebase to another. This idea is illustrated in Figure 3.4. Nodes P1 and P2 send out synchronization beacons, and they create two overlapping neighborhoods, with node B lying in the overlapping area. Since node A and node B lie within the same neighborhood, their clock relationship (i.e., clock offset and skew) can be established from node P1's reference broadcast. Similarly the clock relationship between node B and node C can be established from node P2's reference broadcast. Therefore, the clock relationship between node A and node C can be computed with node B acting as a gateway.

3.2.4 Extension of FTSP

FTSP can be extended to network-wide synchronization in a straightforward manner. First, a root node, to which the whole network is being synchronized, is elected by the network. Nodes that are within the broadcast radius of the root node can receive timestamped messages from the root node. They then estimate the offset and skew of their own local clocks, thus synchronizing with the root node. The newly synchronized nodes can then broadcast synchronization messages to other nodes in the network. The advantage of this flooding process is that it begins with the root node, and there is no need to have a level hierarchy, unlike with TPSN.

3.2.5 Pairwise Broadcast Synchronization (PBS)

The PBS protocol employs both sender–receiver and receiver-only synchronization approaches to achieve network-wide synchronization with high energy efficiency [66], [68]. In PBS, a number of sensor nodes can be synchronized by only overhearing timing messages being exchanged between pairs of nodes. Such a signaling mechanism significantly reduces the overall energy consumption by decreasing the number of timing messages required in synchronization. PBS requires a much smaller number of timing messages than other protocols such as RBS, TPSN, and FTSP, and its benefits greatly increase as the sensors are more densely deployed.

In fact, a similar concept combining the merits of both RRS and SRS approaches has been applied in the time diffusion protocol (TDP) [53]. TDP elects diffusion leaders in every level of the network and the selected leaders successively broadcast synchronization messages. However, unlike TDP, the PBS protocol selects the best set of synchronization pairs to minimize the overall number of timing messages and energy consumption, while TDP is based on very different election criteria: the balance of work loads and the clock stability. Moreover, in TDP, there are no concerns about the optimum number of diffusion leaders in terms of energy efficiency and how to guarantee the network-wide synchronization. Chapter 8 illustrates and analyzes the PBS protocol in detail.

3.2.6 Time Diffusion Protocol (TDP)

TDP [90] is a protocol enabling the sensor network to reach an equilibrium time with the clocks of individual sensors within a small time deviation from the equilibrium time. The protocol can be understood as periodically applying the following three phases: (1) election of master/diffused leader nodes, (2) the time diffusion procedure, and (3) the peer evaluation procedure. It is shown analytically in [90] that TDP enables the clocks in the whole network to converge to a unique value.

In the first phase, master nodes are elected in the sensor field. The election criteria include the quality of the clock and the energy resources of a particular node. Referring to Figure 3.5, assume that node P is elected to be the master node (here we illustrate the concept with one master node, while in more complicated networks, more than one master node might be possible). Node P then sends a number of timestamped messages to its neighbors. Once the neighbors receive the messages, they self-determine whether to become diffused leader nodes, based on the

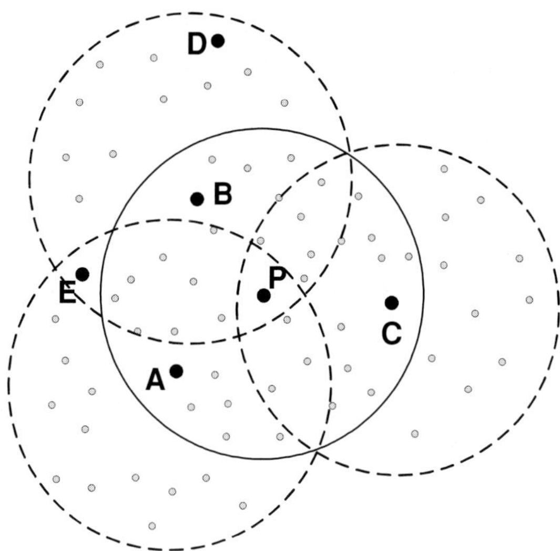

Figure 3.5: Time-diffusion synchronization protocol.

results of the last round peer evaluation procedure (the third phase). In Figure 3.5, nodes A, B, and C are the elected diffused leader nodes. The elected diffused leader nodes respond to the master node, thus enabling the master node to measure the average and the standard deviation of the round-trip delay from its neighbors. At the same time, the diffused leader nodes start sending messages to their own neighbors to measure the mean and standard deviation of round-trip delays to their neighbors. The process is repeated until all the nodes have been covered.

In the second phase, the time information from the master node is diffused (with the help of diffused leader nodes) to all the nodes in the network. The diffusion procedure takes place according to the following sequence of events. First, the master node sends a timestamped message containing the standard deviation of the round-trip delay to its neighbors. Before transmission, the timestamp of the message is adjusted by half of the measured average round-trip delay (from the first phase) to account for the message delivery delay to its neighbors. Once the diffused leader nodes receive the timestamped message, they set their clocks according to the received timestamp and then broadcast their own timestamped messages, containing their measured standard derivations of the round-trip times to their neighbors. Again, before transmission of the messages, the timestamps have to be adjusted by half of the measured average round-trip delays to their neighbors. For nodes that are not diffused leaders, if they only receive a message from one diffused leader node (e.g., node D in Figure 3.5), they just set their clock according to the timestamp they received. Nodes that have received more than one timestamped message originating from different diffused leader nodes (e.g., node E in Figure 3.5) use the standard deviations as weightings (the smaller the deviation, the larger the weighting) to combine the clock values and set their clocks according to the result.

The purpose of the third phase is to allow the sensor nodes to evaluate the stability of their local clocks. First, the elected master nodes broadcast a number of timestamped messages. The neighbor nodes receiving these messages calculate the 2-sample Allan[1] variance [90] of the local clock from the clock of the master nodes and send back these calculated Allan variances to the master nodes. Then the master nodes compute the average of all the Allan variances they received and send the result back to their neighbor nodes [3]. By this procedure, all the neighbor nodes can evaluate the quality of their clocks with respect to those of their neighbors by comparing their calculated Allan variance with the average value. The above procedure is repeated, but with the elected diffused leader nodes broadcasting the timestamped messages.

3.2.7 Synchronous and Asynchronous Diffusion Algorithms

In [53], two diffusion algorithms are proposed. One is called the rate-based synchronous diffusion algorithm. The idea behind this algorithm is that in order for a network to achieve an equilibrium time, the clock at node i, denoted c_i, should be adjusted according to the differences between its clock and its neighbors' clocks (assuming node i has exchanged clock readings with its neighbors). That is, the clock at node i should be set to $c_i - \sum_{j \neq i} r_{ij}(c_i - c_j)$, where $r_{ij} > 0$ is the diffusion rate, $r_{ij} = 0$ if node i and node j cannot directly communicate and the condition $\sum_{j \neq i} r_{ij} \leq 1$ is enforced. The above algorithm can also be formulated using matrix notation. For a group of n sensor nodes, let $\mathbf{c}^t$ be the vector of length n containing the clock readings of all the sensor nodes at time t. The synchronous diffusion algorithm adjusts the clocks of different nodes via the recursion $\mathbf{c}^{t+1} = \mathbf{R}\mathbf{c}^t$, where the transition matrix $\mathbf{R}$ assumes the expression

$$\mathbf{R} = \begin{pmatrix} r_{11} & r_{12} & \cdots & r_{1n} \\ r_{21} & r_{22} & \cdots & r_{2n} \\ \vdots & \vdots & \ddots & \vdots \\ r_{n1} & r_{n2} & \cdots & r_{nn} \end{pmatrix} \tag{3.10}$$

and $r_{ii} = 1 - \sum_{j \neq i} r_{ij}$. It is shown in [53] that if the second largest eigenvalue of $\mathbf{R}$ is smaller than 1, the synchronous diffusion algorithm will converge, in the sense that all the elements in $\mathbf{c}^t$ will become equal asymptotically as time t increases.

The synchronous diffusion algorithm requires all the nodes to operate in an ordered manner. In order to remove this constraint, Li and Rus [53] proposed another

[1]Allan variance is a measurement of stability in clocks and oscillators.

algorithm, called the asynchronous diffusion algorithm. In this algorithm, each node asks its neighbors about their clock readings and computes an average or median value based on those clock readings. Then the average value is sent back to the neighbors so they can update their clocks. This algorithm yields a very simple averaging operation of a node over its neighbors' clocks and the averaging operations by different nodes can be carried out at different times and in any order. Therefore, the term asynchronous is associated with this algorithm. It is shown in [53] that the clocks of sensor nodes in a sensor network converge to the average value by using this asynchronous algorithm.

3.2.8 Protocols Based on Pulse Transmissions

Synchronization schemes that operate exclusively at the physical layer by transmitting pulses instead of message packets have been proposed in [36] and [40]. In [36], inspired by the synchronously flashing fireflies, the time synchronization problem in sensor networks is modeled using pulse coupled oscillators (PCOs). In this scheme, each node (say node j) in the sensor network is associated with an increasing monotonic state function $x_j(t)$ taking values from 0 to 1. If a node is isolated, the state function $x_j(t)$ increases from 0 to 1 smoothly as a function of time and the node emits a pulse when the state function achieves the unit value ($x_j(t) = 1$). After firing a pulse, the node immediately resets its state to zero. This results in the periodic emission of pulses with period T. If a node is not isolated, it can receive pulses from other nodes. When a node receives a pulse, its state variable changes as follows:

$$x_j(\Delta^+) = \begin{cases} x_j(\Delta) + \xi, & \text{if } x_j(\Delta) + \xi < 1 \\ 0, & \text{otherwise} \end{cases}, \tag{3.11}$$

where Δ is the time taken for the pulse to be received and ξ is the advancement of the clock phase. This means that a node receiving a pulse either emits the pulse at the same time or shortens the waiting time for the next round of emissions, with the assumption that after a node fires a pulse, it enters a short refractory period, during which no signal can be received from other nodes (to avoid infinite feedback). It can be shown that only when the nodes emit the pulse simultaneously will they be insensitive to coupling, and therefore achieve synchronization.

In [40], a cooperative technique that constructs a sequence of pulses with equidistant zero-crossings is developed. The basic idea of this scheme is as follows. Assume that there is a leader node and that it emits a sequence of pulses with equidistant zero-crossings. The surrounding nodes receive this pulse sequence, and based on the locations of the observed zero-crossings, the surrounding nodes predict when the next pulse will be transmitted. Then, these nodes emit pulses at their predicted times and an aggregate pulse sequence is generated. It is shown in [40] that although the prediction at an individual node may not be perfect, under certain conditions on the pulse and in asymptotically dense networks, the zero-crossings of the aggregate waveform sequence will be at the same positions as the zero-crossings of the original waveform sequence emitted by the leader node due to spatial averaging. This aggregate pulse sequence is heard by the nodes lying further away from the leader node and these nodes perform prediction as described above and emit their pulses at their predicted times. The procedure continues until all the nodes are synchronized.

Notice that the synchronization algorithms discussed in this subsection only provide a unified ticking rhythm across sensor nodes, not the synchronization of clock time. A good analogy is a group of people clapping together to get a rhythm. However, there are applications in which a unified rhythm is enough, e.g., in distributed beamforming and the reachback communication channel [39]. Another variation, a joint physical- and network-layer time synchronization scheme, was proposed in [94] to overcome the effects of imperfect physical-layer synchronization due to the nature of common wireless channels.

3.3 ADAPTIVE TIME SYNCHRONIZATION

While all the protocols mentioned in this chapter can achieve instantaneous synchronization among nodes, the timing of different nodes drifts apart as time passes. Therefore, periodic resynchronization is needed to maintain long-term synchronization. Intuitively, less frequent resynchronization requires less energy but leads to a larger synchronization error, while more frequent resynchronization leads to a smaller synchronization error but requires more energy. A natural question is what is the minimum resynchronization frequency (or equivalently maximum resynchronization period) that can meet the desired synchronization precision. These considerations motivate the need for adaptive algorithms to dynamically determine the resynchronization period, the number of beacons to be used in each round of synchronization, synchronization accuracy, and so on. In this section, we will review three adaptive time synchronization algorithms proposed in the literature.

3.3.1 RATE-ADAPTIVE TIME SYNCHRONIZATION (RATS)

Consider the case where node A sends timestamped messages to node B periodically with period τ, and node B records the receiving times of the messages. Based on a number of data points $(T_i^{(A)}, T_i^{(B)})$, where $T_i^{(A)}$ and $T_i^{(B)}$ are the timestamps made at node A and node B, respectively, node B determines the largest τ such that the synchronization error is smaller than a certain limit. The RATS [28] determines the optimal τ in an iterative or recursive manner. The idea behind it can be

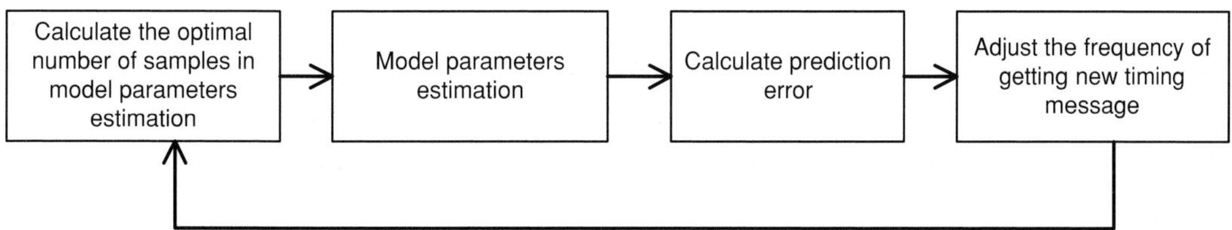

Figure 3.6: Flow chart of RATS run at the node receiving timestamped messages from another node.

summarized using the flow chart shown in Figure 3.6. First, node B calculates the optimal number of data samples for model parameters (e.g., clock offset and skew) estimation based on the current value of τ. Next, node B takes the required number of data points (stored in memory) and estimates the model parameters. Then, node B computes the prediction error. Finally, using the calculated prediction error, node B adjusts the frequency of getting a new timing message from node A: if the prediction error is larger than the upper limit threshold E_u, it means that the timing message rate from node A is not frequent enough, therefore τ should be decreased. On the other hand, if the prediction error is smaller than the lower limit threshold E_l, that translates into fewer timing messages, thus τ should be increased. Multiplicative increase and decrease strategies are used to enable fast convergence and quick response to the changing environment. After getting a new data point according to the new value of τ, the above process is repeated.

3.3.2 RBS-Based Adaptive Clock Synchronization

Within the RBS setting, PalChaudhuri *et al.* [69] extend the deterministic RBS protocol (discussed in Subsection 3.1.3) to an adaptive probabilistic synchronization algorithm, allowing tradeoffs between synchronization accuracy and resource expenditure. This algorithm is based on the observation that if the relative clock skew error between two nodes ε is a Gaussian RV with zero mean and variance σ^2, then the probability of error-free synchronization with N broadcasting messages is given by

$$Pr(|\varepsilon| < \varepsilon_{max}) = 2\text{erf}\left(\frac{\sqrt{N}\varepsilon_{max}}{\sigma}\right), \tag{3.12}$$

where ε_{max} stands for the maximum specified (allowable) clock offset for communications, and the error function erf is defined via $\text{erf}(x) \triangleq (1/2\pi)\int_0^x \exp\left(-t^2/2\right)dt$. From the above equation, it is clear that the performance criterion is a probabilistic measure since there is always a possibility that the clock offset is greater than some limit ε_{max}. However, one can reduce this probability to an arbitrarily small value by increasing N, the number of broadcasting messages in one round of RBS.

After application of RBS, we can bound the clock skew error with a certain probability. However, since clocks from different nodes would drift apart as time passes, we need to reapply RBS periodically. In [69] PalChaudhuri *et al.* propose using the following formula to determine the maximum time between resynchronizations τ_{max}

$$\tau_{max} = \frac{\gamma_{max} - \varepsilon_{max}}{\rho} - d_{max}, \tag{3.13}$$

where γ_{max} denotes the maximum allowable clock skew at any time, ρ denotes the maximum drift of clock, and d_{max} is the maximum delay of timestamp exchanges in RBS. With different synchronization precision requirements (specified by γ_{max}), one can determine the required resynchronization period τ_{max}.

3.3.3 ADAPTIVE MULTI-HOP TIME SYNCHRONIZATION (AMTS)

The AMTS protocol is based on a similar system model to TPSN and employs a number of novel features as well [66], [68]. It consists of three functional phases: the *network level discovery phase*, the *synchronization phase*, and the *network evaluation phase*, and a number of network parameters such as the latency factor, the average number of hops, and the resynchronization period to optimize the synchronization protocol. Relative to TPSN, AMTS assumes the additional *network evaluation phase*, while the functions of the other two phases are similar to the ones encountered in TPSN.

Robustness to high latencies and network delays is ensured based on the clock estimators presented in this book, and therefore AMTS fits well for sensor network applications presenting large delays in timing message exchanges such as underwater acoustic sensor networks [93]. In addition, AMTS adapts the joint clock offset and skew estimators to increase the resynchronization period.

As in TPSN, generating a hierarchical structure in the network, the level discovery phase, is the first step of AMTS. In this phase, every single node in the network is assigned a level and prepares for synchronization. The second step of AMTS, called the time synchronization phase, consists of pairwise synchronizations between adjacent nodes until every node in the network is synchronized to the reference. In the synchronization phase, AMTS estimates not only the current clock offset but also the clock frequency (skew) to guarantee long-term reliability of synchronization, while TPSN only estimates the clock offset. Hence, AMTS requires far less frequent resynchronizations. Finally, the reference node investigates the current status of network traffic in order to optimize the resynchronization period and the number of beacons in terms of energy efficiency. Moreover, it selects the synchronization mode to be either *always on* (AO) (always maintain network-wide synchronization) or *sensor initiated* (SI) (synchronize only when it needs to) based on the network status. This step is the network evaluation phase, and its goal is to minimize the number of message exchanges for synchronization in a given time, i.e., it aims to minimize the total energy consumption for synchronization. AMTS periodically repeats the synchronization and network evaluation phases to minimize the total energy consumption with respect to the current network status. Chapter 11 describes these procedures in detail.

Chapter 4

FUNDAMENTAL APPROACHES TO TIME SYNCHRONIZATION

As described in Chapter 3, various protocols targeting clock synchronization in WSNs have been proposed, mainly based on packet synchronization techniques. In general, this family of protocols can be broadly divided into two fundamental approaches: sender–receiver synchronization (SRS), see, e.g., [29], [53], [74], [79], [86], and receiver–receiver synchronization (RRS), see, e.g., [22], [63], [69], [90]. SRS relies on the traditional model of two-way message exchanges between a pair of nodes. For RRS, the nodes to be synchronized first receive a beacon packet from a common sender, then compare the receiving times of the beacon packet to compute the relative clock offsets. Most of the existing time synchronization protocols rely on one of these two approaches. For instance, NTP [61] and TPSN [29] adopt SRS since they depend on a series of pairwise synchronizations that assume two-way timing message exchanges. Notice also that the RBS protocol [22] relies on RRS since it requires pairs of message exchanges among children nodes (except the reference) to compensate their relative clock offsets.

A new approach for time synchronization, called receiver-only synchronization (ROS) [68], has also been proposed. The aim of ROS is to minimize the number of required timing messages and energy consumption during synchronization while preserving a high level of accuracy [68]. This approach can be used to achieve network-wide synchronization with many fewer timing messages than other well-known protocols such as TPSN and RBS.

Next we will present and analyze each of these synchronization approaches and illustrate how the general design considerations can be resolved in these approaches. For all these approaches, we only present the underlying signaling mechanisms

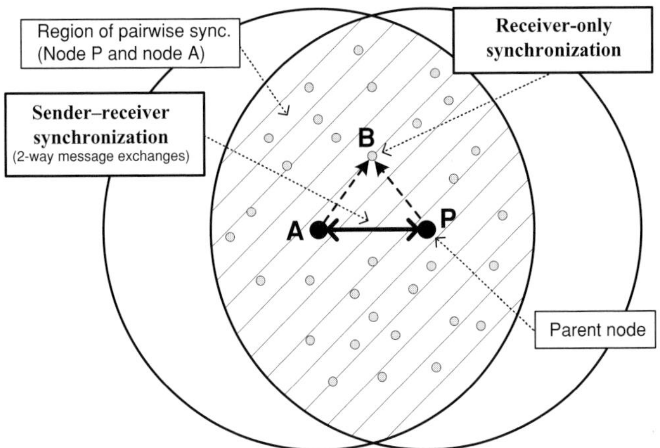

Figure 4.1: Sender–receiver synchronization and receiver-only synchronization.

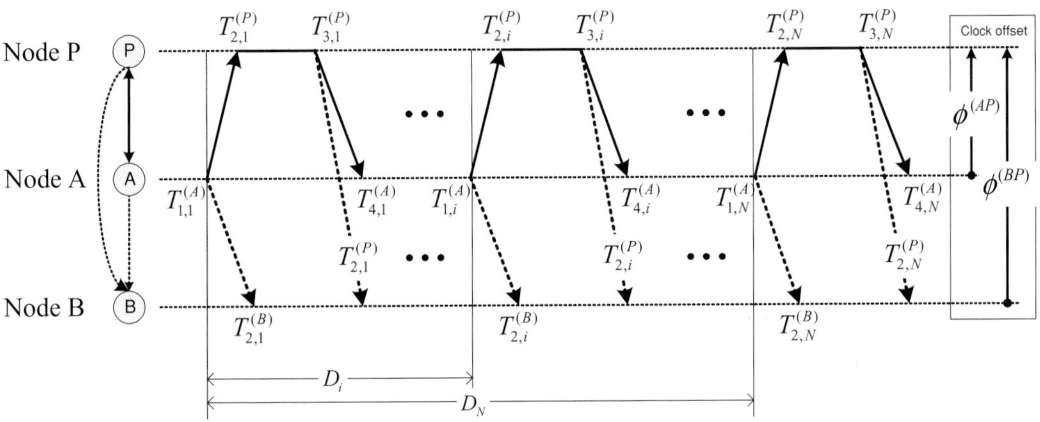

Figure 4.2: Clock synchronization model of SRS and ROS.

for performing pairwise synchronization, i.e., synchronizing a pair of nodes, since network-wide synchronization can be simply achieved in various ways, essentially by performing multiple groups of pairwise synchronizations.

4.1 SENDER–RECEIVER SYNCHRONIZATION (SRS)

This approach is based on the classical two-way timing message exchange mechanism between two adjacent nodes. Consider a parent node P and one of its child node, node A as is depicted in Figure 4.1. The clock model for the two-way message exchange is depicted in Figure 4.2, where $\phi^{(AP)}$ denotes the clock offset between node A and node P, and the timing messages are assumed to be exchanged multiple

(N) times [29], [91]. Here, the timestamps made during the ith message exchange $T_{1,i}^{(A)}$ and $T_{4,i}^{(A)}$ are measured by the local clock of node A, and $T_{2,i}^{(P)}$ and $T_{3,i}^{(P)}$ are measured by the local clock of node P. Node A transmits a synchronization packet, containing the value of timestamp $T_{1,i}^{(A)}$ to node P. Node P receives it at time $T_{2,i}^{(P)}$ and transmits an acknowledgement packet to node A at $T_{3,i}^{(P)}$. This packet contains the value of timestamps $T_{1,i}^{(A)}$, $T_{2,i}^{(P)}$, and $T_{3,i}^{(P)}$. Then, node A finally receives the packet at $T_{4,i}^{(A)}$.

As discussed before, packet delays can be split into several different components: send, access, transmission, propagation, and receive times. These delay components are embedded into two major constituent elements: the fixed portion d and the variable portion X_i. The variable portion of delays depends on various network parameters (e.g., network status, traffic, etc.) and setup variables, and therefore no single delay model can be found to fit for every case [49]. Thus far, several probability density function (pdf) models have been proposed for modeling random delays, the most widely deployed ones being Gaussian, Gamma, exponential and Weibull pdfs [22], [52], and [70]. The Gaussian delay model is appropriate if the delays are regarded as numerous independent random processes. In [22], the chi-square test showed that the variable portion of delays can be modeled as Gaussian distributed RVs with 99.8% confidence. On the other hand, a single-server M/M/1 queue can fittingly represent the cumulative link delay for point-to-point hypothetical reference connection, where the random delays are independently modeled as exponential RVs [1]. Thus, we assume the random portions of delays are either normally or exponentially distributed RVs.

Suppose that the clock frequencies of two nodes remain equal during the synchronization period, and both $X_i^{(AP)}$ and $X_i^{(PA)}$ are normally distributed RVs with mean μ and variance $\sigma^2/2$. From Figure 4.2, $T_{2,i}^{(P)}$ and $T_{4,i}^{(A)}$ can be expressed as

$$T_{2,i}^{(P)} = T_{1,i}^{(A)} + \phi^{(AP)} + d^{(AP)} + X_i^{(AP)}, \tag{4.1}$$

$$T_{4,i}^{(A)} = T_{3,i}^{(P)} + \phi^{(PA)} + d^{(PA)} + X_i^{(PA)}, \tag{4.2}$$

where $\phi^{(PA)} = -\phi^{(AP)}$, and $d^{(AP)}$ and $X_i^{(AP)}$ denote the fixed and random portions of timing delays in the message transmissions from node A to node P, respectively. By defining the delays in the uplink $U_i \triangleq T_{2,i}^{(P)} - T_{1,i}^{(A)}$ and the downlink $V_i \triangleq T_{4,i}^{(A)} - T_{3,i}^{(P)}$, the ith delay observations for the uplink and the downlink corresponding to the ith timing message exchange are given by $U_i = \phi^{(AP)} + d^{(AP)} + X_i^{(AP)}$ and $V_i = \phi^{(PA)} + d^{(PA)} + X_i^{(PA)}$, respectively. Then, the likelihood function based on the observations $\{U_i\}_{i=1}^N$ and $\{V_i\}_{i=1}^N$ is given by

$$L(\phi^{(AP)}, \mu, \sigma^2) = (\pi\sigma^2)^{-\frac{N}{2}} \exp\left\{-\frac{1}{\sigma^2}\left[\sum_{i=1}^{N}(U_i - d^{(AP)} - \phi^{(AP)} - \mu)^2 + \sum_{i=1}^{N}(V_i - d^{(PA)} + \phi^{(AP)} - \mu)^2\right]\right\},$$

where N stands for the number of message exchanges. Differentiating the log-likelihood function leads to

$$\frac{\partial \ln L(\phi^{(AP)})}{\partial \phi^{(AP)}} = -\frac{2}{\sigma^2}\sum_{i=1}^{N}[\phi^{(AP)} + d^{(AP)} - d^{(PA)} - (U_i - V_i)].$$

The fixed portions of delays are mainly determined by the propagation delays, and both uplink and downlink channels assume the same propagation distance. Thus, the fixed portions of delays $d^{(AP)}$ and $d^{(PA)}$ are assumed to be equal, and are denoted by d for the rest of this chapter. Indeed, the propagation delay is less than 1 microsecond for ranges under 300 meters, hence is almost negligible when compared to other dominant delay components which have times of the order of hundreds of milliseconds [58]. The MLE of clock offset is given by

$$\hat{\phi}^{(AP)} = \arg\max_{\phi^{(AP)}}[\ln L(\phi^{(AP)})] = \frac{\overline{U} - \overline{V}}{2}. \tag{4.3}$$

Thus, node A can be synchronized to the parent node P by simply taking the difference of the average delay observations $\overline{U}$ and $\overline{V}$.

For exponential random delays $X_i^{(PA)}$ and $X_i^{(AP)}$ with the same mean λ, the likelihood function based on the observations $\{U_i\}_{i=1}^N$ and $\{V_i\}_{i=1}^N$ becomes

$$L(\phi^{(\mathrm{AP})}, \lambda) = \lambda^{-2N} e^{-\frac{1}{\lambda}\sum_{i=1}^{N}[U_i + V_i - 2d]} \prod_{i=1}^{N} I\left[U_i - \phi^{(\mathrm{AP})} - d \geq 0,\ V_i + \phi^{(\mathrm{AP})} - d \geq 0\right],$$

where $I(\cdot)$ stands for the indicator function (i.e., $I(\cdot)$ is 1 whenever its inner condition holds, otherwise it is 0). In [41], Jeske proved that the maximum likelihood estimator of $\phi^{(AP)}$ exists when d is unknown and exhibits the same form as the estimator proposed in [71] and [72], namely

$$\hat{\phi}^{(AP)} = \frac{\min_{1\leq i\leq N} U_i - \min_{1\leq i\leq N} V_i}{2}, \tag{4.4}$$

From (4.3) and (4.4), it is clear that if only one round of message exchanges is performed ($N = 1$), the MLEs of clock offset for exponential and Gaussian delay models coincide and assume the expression $\hat{\phi}^{(AP)} = (U - V)/2$, which is exactly the same clock offset estimator adopted in TPSN [29] and NTP [61]. However, as soon as $N > 1$, the expressions for MLEs corresponding to the clock offset in Gaussian and exponential network delay become different.

Note that the clock offset between two nodes generally keeps increasing due to the difference in the clock parameters of each oscillator. Therefore, applying a clock skew correction mechanism increases the synchronization accuracy and guarantees the long-term reliability of synchronization. Chapter 6 describes the joint clock offset and skew estimators for the SRS approach. In addition, a family of robust and practical clock offset and skew estimators which do not require prior knowledge of d is proposed in Chapter 7.

4.2 RECEIVER-ONLY SYNCHRONIZATION (ROS)

Due to the power constraint, the communication range of a sensor is strictly limited to a (radio-geometrical) circle whose radius depends on the transmission power (see Figure 4.1). In this figure, every node within the hatched area (e.g., node B) can receive messages from both node P and node A. Suppose that node P is a parent (or reference) node, and node P and node A perform a pairwise synchronization using two-way timing message exchanges as depicted in Figure 4.2. Then, all the nodes in the common coverage region of node P and node A (the hatched region) receive a series of synchronization messages containing the information about the timestamps of the pairwise synchronization. Using this information, node B can be also synchronized to the parent node P with no extra timing message transmissions. This approach is called receiver-only synchronization (ROS). In general, all the sensor nodes lying within the hatched area in Figure 4.1 can be synchronized by only receiving timing messages through the ROS approach. Here, node P and node A can be regarded as supernodes since they provide synchronization beacons for all the nodes located in their vicinity.

In Figure 4.1, consider an arbitrary node, say node B, in the hatched region. While node P and node A exchange time messages, node B can overhear these time messages. Hence, node B is able to observe a set of time readings ($\{T_{2,i}^{(B)}\}_{i=1}^{N}$) at its local clock when it receives packets from node A as depicted in Figure 4.2. In addition, node B can also receive the information about a set of timestamps $\{T_{2,i}^{(P)}\}_{i=1}^{N}$ by receiving the packets transmitted by node P. Considering the effects of both clock offset and skew, the reception time at node P in the ith up-link message $T_{2,i}^{(P)}$ is given by

$$T_{2,i}^{(P)} = T_{1,i}^{(A)} + \phi^{(AP)} + \omega^{(AP)}(T_{1,i}^{(A)} - T_{1,1}^{(A)}) + d^{(AP)} + X_i^{(AP)}, \qquad (4.5)$$

where $\omega^{(AP)}$ stands for the relative clock skew between node A and node P. Likewise, the reception time at node B in the ith uplink message $T_{2,i}^{(B)}$ can be represented by

$$T_{2,i}^{(B)} = T_{1,i}^{(A)} + \phi^{(AB)} + \omega^{(AB)}(T_{1,i}^{(A)} - T_{1,1}^{(A)}) + d^{(AB)} + X_i^{(AB)}, \tag{4.6}$$

where $\phi^{(AB)}$ and $\omega^{(AB)}$ stand for the relative clock offset and skew between node A and node B, $d^{(AB)}$ and $X_i^{(AB)}$ denote the fixed and random portions of timing delays in the message transmission from node A to node B, respectively. Here, $X_i^{(AB)}$ is assumed to be a normally distributed RV with mean μ and variance $\sigma^2/2$.

The linear regression technique can be applied to synchronize node B and compensate the effects of the relative clock skew between node P and node B. Subtracting (4.6) from (4.5) yields

$$T_{2,i}^{(P)} - T_{2,i}^{(B)} = \phi^{(BP)} + \omega^{(BP)}(T_{1,i}^{(A)} - T_{1,1}^{(A)}) + d^{(AP)} - d^{(AB)} + X_i^{(AP)} - X_i^{(AB)}. \tag{4.7}$$

Since $d^{(AB)}$ and $d^{(AP)}$ are fixed values and $X_i^{(AB)}$ and $X_i^{(AP)}$ are normally distributed RVs, the noise component can be defined in (4.7) by $z[i] \triangleq \mu' + X_i^{(AP)} - X_i^{(AB)}$, where $\mu' \triangleq d^{(AP)} - d^{(AB)}$ and $z[i]$ assumes the normal distribution $z[i] \sim \mathcal{N}(\mu', \sigma^2)$. Let $x[i] \triangleq T_{2,i}^{(P)} - T_{2,i}^{(B)} - \mu'$ and $w[i] \triangleq z[i] - \mu'$, then the set of observed data in (4.7) can be written in matrix notation as follows:

$$\mathbf{x} = \mathbf{H}\boldsymbol{\theta} + \mathbf{w},$$

where $\mathbf{x} = [x[1]\ x[2]\ \cdots\ x[N]]^T$, $\mathbf{w} = [w[1]\ w[2]\ \cdots\ w[N]]^T$, $\boldsymbol{\theta} = [\phi^{(BP)}\ \omega^{(BP)}]^T$, and

$$\mathbf{H} = \begin{bmatrix} 1 & 1 & \cdots & 1 \\ 0 & T_{1,2}^{(A)} - T_{1,1}^{(A)} & \cdots & T_{1,N}^{(A)} - T_{1,1}^{(A)} \end{bmatrix}^T.$$

Note that the noise vector $\mathbf{w} \sim \mathcal{N}(0, \sigma^2\mathbf{I})$ and the observation matrix $\mathbf{H}$ assumes the dimension $N \times 2$. From [44, Theorem 3.2, p. 44], the minimum variance unbiased (MVU) estimator for the relative clock offset and skew is given by $\hat{\boldsymbol{\theta}} = \mathbf{g}(\mathbf{x})$, where $\mathbf{g}(\mathbf{x})$ satisfies

$$\frac{\partial \ln p(\mathbf{x}; \boldsymbol{\theta})}{\partial \boldsymbol{\theta}} = \mathbf{I}(\boldsymbol{\theta})(\mathbf{g}(\mathbf{x}) - \boldsymbol{\theta}). \tag{4.8}$$

Since the noise vector $\mathbf{w}$ is zero mean and Gaussian distributed, from the results in [44, p. 85], the derivative of the log-likelihood function can be expressed as

$$\frac{\partial \ln p(\mathbf{x}; \boldsymbol{\theta})}{\partial \boldsymbol{\theta}} = \frac{\mathbf{H}^T\mathbf{H}}{\sigma^2}[(\mathbf{H}^T\mathbf{H})^{-1}\mathbf{H}^T\mathbf{x} - \boldsymbol{\theta}], \tag{4.9}$$

where $\mathbf{H}^T\mathbf{H}$ is assumed to be invertible. Therefore, comparing (4.8) with (4.9) yields

$$\hat{\boldsymbol{\theta}} = (\mathbf{H}^T\mathbf{H})^{-1}\mathbf{H}^T\mathbf{x}, \tag{4.10}$$

$$\mathbf{I}(\boldsymbol{\theta}) = \frac{\mathbf{H}^T\mathbf{H}}{\sigma^2}, \tag{4.11}$$

where $\mathbf{I}(\boldsymbol{\theta})$ is the Fisher information matrix. After some mathematical manipulations, the joint clock offset and skew estimator can be expressed as

$$\begin{bmatrix} \hat{\phi}^{(BP)} \\ \hat{\omega}^{(BP)} \end{bmatrix} = \frac{1}{N\sum\limits_{i=1}^{N} D_i^2 - \left[\sum\limits_{i=1}^{N} D_i\right]^2} \begin{bmatrix} \sum\limits_{i=1}^{N} D_i^2 \sum\limits_{i=1}^{N} x[i] - \sum\limits_{i=1}^{N} D_i \sum\limits_{i=1}^{N} [D_i x[i]] \\ N\sum\limits_{i=1}^{N} [D_i x[i]] - \sum\limits_{i=1}^{N} D_i \sum\limits_{i=1}^{N} x[i] \end{bmatrix}, \tag{4.12}$$

where $D_i \triangleq T_{1,i}^{(A)} - T_{1,1}^{(A)}$. The CRLB can be obtained by inverting the Fisher information matrix $\mathbf{I}(\boldsymbol{\theta})$. From (4.11), the Fisher information matrix is given by

$$\mathbf{I}(\boldsymbol{\theta}) = \frac{1}{\sigma^2} \begin{bmatrix} N & \sum\limits_{i=1}^{N} D_i \\ \sum\limits_{i=1}^{N} D_i & \sum\limits_{i=1}^{N} D_i^2 \end{bmatrix}.$$

Then, inverting $\mathbf{I}(\boldsymbol{\theta})$ yields

$$\mathbf{I}^{-1}(\boldsymbol{\theta}) = \frac{\sigma^2}{N\sum\limits_{i=1}^{N} D_i^2 - \left[\sum\limits_{i=1}^{N} D_i\right]^2} \begin{bmatrix} \sum\limits_{i=1}^{N} D_i^2 & -\sum\limits_{i=1}^{N} D_i \\ -\sum\limits_{i=1}^{N} D_i & N \end{bmatrix}. \tag{4.13}$$

Hence, from (4.13), the Cramer–Rao Bounds (CRBs) for the relative clock offset and skew become

$$\mathrm{var}(\hat{\phi}^{(BP)}) \geq \frac{\sigma^2 \sum\limits_{i=1}^{N} D_i^2}{N\sum\limits_{i=1}^{N} D_i^2 - \left[\sum\limits_{i=1}^{N} D_i\right]^2} \tag{4.14}$$

and

$$\mathrm{var}(\hat{\omega}^{(BP)}) \geq \frac{\sigma^2 N}{N\sum\limits_{i=1}^{N} D_i^2 - \left[\sum\limits_{i=1}^{N} D_i\right]^2}. \tag{4.15}$$

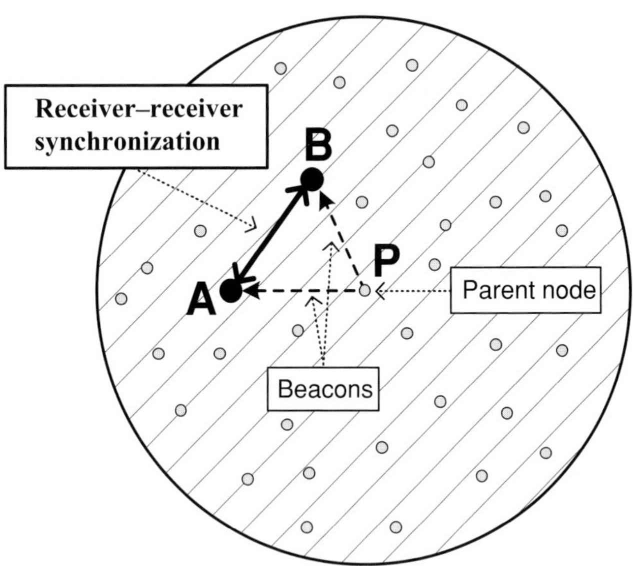

Figure 4.3: Receiver–receiver synchronization.

Notice also that the regularity conditions for the CRBs hold:

$$
\begin{aligned}
E\left[\frac{\partial \ln p(\mathbf{x};\boldsymbol{\theta})}{\partial \boldsymbol{\theta}}\right] &= \begin{bmatrix} E\left[\frac{\partial \ln p(\mathbf{x};\boldsymbol{\theta})}{\partial \phi^{(BP)}}\right] \\ E\left[\frac{\partial \ln p(\mathbf{x};\boldsymbol{\theta})}{\partial \omega^{(BP)}}\right] \end{bmatrix} \\
&= \begin{bmatrix} E\left[\frac{1}{\sigma^2}\sum_{i=1}^{N}\left[x[n]-\phi^{(BP)}-\omega^{(BP)}D_i\right]\right] \\ E\left[\frac{1}{\sigma^2}\sum_{i=1}^{N}\left\{\left[x[n]-\phi^{(BP)}-\omega^{(BP)}D_i\right]D_i\right\}\right] \end{bmatrix} = \mathbf{0}.
\end{aligned}
$$

Consequently, using the results in (4.12), node B can be synchronized with node P. Likewise, all the other nodes in the hatched region in Figure 4.1 can be simultaneously synchronized to the parent node, node P, without any additional timing message transmissions, thus saving a significant amount of energy.

4.3 Receiver–Receiver Synchronization (RRS)

RRS is an approach for synchronizing a set of child nodes that receive the beacon messages from a common sender (a reference or parent node). Consider a parent (reference) node P and arbitrary nodes A and B, which are located within the communication range of the parent node in Figure 4.3. Suppose as depicted in Figure 4.4 that both node A and node B receive the ith beacon from node P at time instants $T_{2,i}^{(A)}$ and $T_{2,i}^{(B)}$ of their local clocks, respectively. Nodes A and B record the arrival

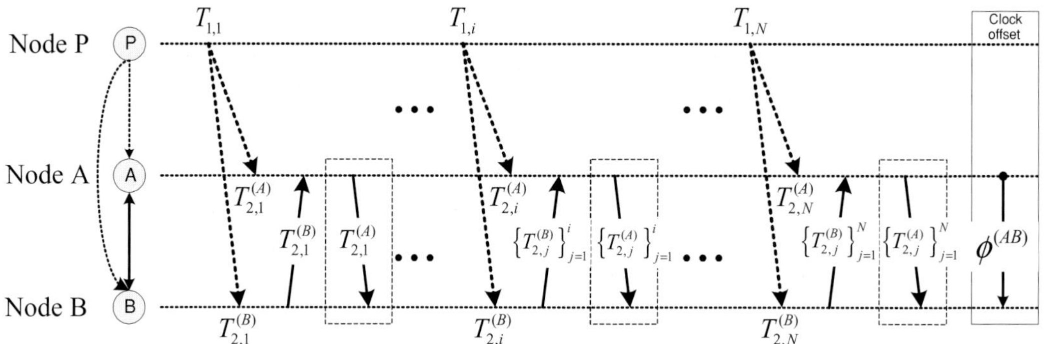

Figure 4.4: Clock synchronization model of RRS.

time of the broadcast packet according to their own timescales and then exchange their timestamps. Suppose $X_i^{(PA)}$ denotes the non-deterministic delay components (random portion of delays) and $d^{(PA)}$ denotes the deterministic delay component (propagation delay) from node P to node A, then $T_{2,i}^{(A)}$ can be written as

$$T_{2,i}^{(A)} = T_{1,i} + d^{(PA)} + X_i^{(PA)} + \phi^{(PA)} + \omega^{(PA)}(T_{1,i} - T_{1,1}), \tag{4.16}$$

where $T_{1,i}$ is the transmission time at the reference node, and $\phi^{(PA)}$ and $\omega^{(PA)}$ are the clock offset and skew of node A with respect to the reference node, respectively. Similarly, we can decompose the arrival time at node B as

$$T_{2,i}^{(B)} = T_{1,i} + d^{(PB)} + X_i^{(PB)} + \phi^{(PB)} + \omega^{(PB)}(T_{1,i} - T_{1,1}), \tag{4.17}$$

where $d^{(PB)}$, $X_i^{(PB)}$, $\phi^{(PB)}$, and $\omega^{(PB)}$ stand for the propagation (fixed) delay, random portion of delays, clock offset and skew of node B with respect to the reference node, respectively.

Subtracting (4.17) from (4.16), we obtain

$$T_{2,i}^{(A)} - T_{2,i}^{(B)} = \phi^{(BA)} + \omega^{(BA)}(T_{1,i} - T_{1,1}) + d^{(PA)} - d^{(PB)} + X_i^{(PA)} - X_i^{(PB)}, \tag{4.18}$$

where $\phi^{(BA)} \triangleq \phi^{(PA)} - \phi^{(PB)}$ and $\omega^{(BA)} \triangleq \omega^{(PA)} - \omega^{(PB)}$ are the relative clock offset and skew between node A and node B at the time they receive the ith broadcast packet from the reference node, respectively. Here, we assume these random portions of delays $X_i^{(PA)}$ and $X_i^{(PB)}$ are normally distributed RVs with mean μ and variance $\sigma^2/2$. Indeed, (4.18) assumes exactly the same form as (4.7). Hence, the same steps can be applied to derive the joint clock offset and skew estimator for ROS. More specifically, define the noise component $z[i] \triangleq \mu' + X_i^{(BA)}$, where

$\mu' \triangleq d^{(PA)} - d^{(PB)}$ and $z[i] \sim \mathcal{N}(\mu', \sigma^2)$. Let us also define $x[i] \triangleq T_{2,i}^{(A)} - T_{2,i}^{(B)} - \mu'$ and $w[i] \triangleq z[i] - \mu'$. Using steps similar to those for ROS, it is straightforward to show, based on (4.18), that the same form of the joint clock offset and skew estimator (4.12) can also be applied to RRS. Consequently, there is no difference between ROS and RRS with regard to the accuracy of synchronization since the effects of random delays are the same. Likewise, the CRB for RRS can also be obtained using a procedure similar to those in ROS. When there is no relative clock skew ($\omega^{(BA)} = 0$), it is straightforward to show that the ML estimator of the relative clock offset $\hat{\phi}^{(BA)}$ becomes

$$\hat{\phi}^{(BA)} = \frac{1}{N} \sum_{i=1}^{N} \left[T_{2,i}^{(A)} - T_{2,i}^{(B)} \right], \tag{4.19}$$

which is the equivalent to the estimator presented in [22].

The main benefit of this approach is that all non-deterministic delay components on the transmitter side (send time and access time) are eliminated. Thus, a high degree of synchronization accuracy can be achieved using this approach.

4.4 Comparisons

SRS can be directly timestamped at the physical layer to eliminate the effects of delay components related to the operating system. Hence, it significantly mitigates the uncertainty of timing delays in message delivery. In contrast, RRS removes the effect of non-deterministic delay components, such as send and access times, on the receiver side. Experimental results using the Berkeley mote platform [29] claim that SRS outperforms RRS in terms of synchronization accuracy (errors) by roughly a factor of 2. However, such an assessment may be disputed since the performance of a synchronization scheme depends on a variety of different factors, such as the network platform and setup, channel status, and estimation scheme.

ROS is designed to minimize the overall energy consumption in synchronization. In this approach, a number of sensor nodes can be synchronized without any message transmission, i.e., they can be synchronized by only receiving timing messages between pairs of nodes. Although there is no gain regarding the synchronization accuracy compared with the other approaches, ROS significantly reduces the overall network-wide energy consumption by decreasing the number of timing messages required for synchronization. Chapter 8 presents additional performance comparisons between different synchronization protocols belonging to one or a mixture of these two general synchronization approaches (SRS and ROS).

Chapter 5

MINIMUM VARIANCE UNBIASED ESTIMATION (MVUE) OF CLOCK OFFSET

Assuming both symmetric and asymmetric exponentially distributed network link delays, this chapter is focused on finding the BLUE-OS and the MVUE for the clock offset between two nodes and evaluates their performance in terms of the MSE, which is chosen as the performance criterion throughout this book. The timing exchange mechanism between the two nodes is the same classical two-way message exchange mechanism adopted in protocols such as TPSN [29], NTP [61], etc.

The main topics in this chapter are as follows. First, BLUE-OS, an estimation scheme that has gone largely unnoticed in engineering literature, is investigated in the context of clock offset and the relevant clock offset estimators are derived. Second, the Rao–Blackwell–Lehmann–Scheffé theorem is used to derive the MVUE and it is shown that the MVUE coincides with the BLUE-OS. Therefore, in the class of unbiased estimators, BLUE-OS is the optimal solution and no other estimator can be found with less MSE (or variance, which is the same as MSE in the unbiased case) than MVUE. For the sake of completeness, the clock offset estimators are also derived for two scenarios, namely when the mean of the exponential link delays is known for each direction and when it is unknown for each direction. Third, a short commentary on whether the MVUE is the best possible solution as compared to the other estimators, such as the MLE, is presented. It is shown that in the most practical scenario, i.e., asymmetric link delays with unknown exponential means, the MLE derived in the presence of symmetric link delays, although biased for asymmetric link delays, outperforms the MVUE in terms of the achievable MSE in the region around the point of link symmetry.

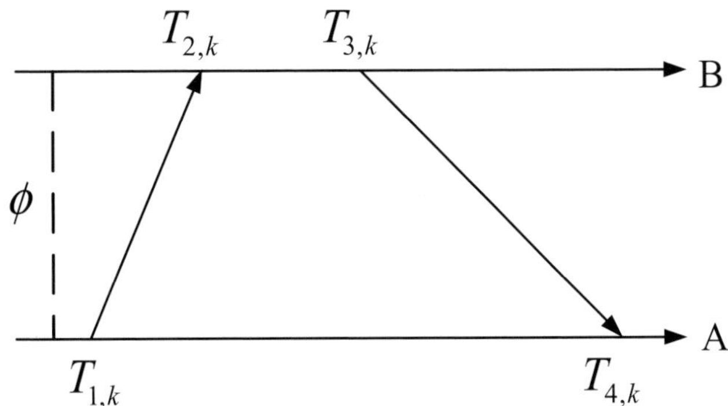

Figure 5.1: A sender–receiver timing message exchange paradigm.

5.1 THE SYSTEM ARCHITECTURE

Adopting the classical approach of sender–receiver synchronization for performing a timing handshake between a pair of nodes, the uplink and downlink timing message exchanges between two clocks A and B are shown in Figure 5.1. The messages $T_{1,k}$ and $T_{4,k}$ represent the times measured by the local clock of node A, while the messages $T_{2,k}$ and $T_{3,k}$ represent the times measured by the local clock of node B (which is also the reference). The synchronization procedure starts at time $T_{1,1}$ and, at each successive message exchange round k, node A sends a synchronization packet containing the timestamp $T_{1,k}$ to node B, which records its reception time as $T_{2,k}$. At $T_{3,k}$, node B sends an acknowledgement packet back to node A containing the timestamps $T_{2,k}$ and $T_{3,k}$, which is delivered and timestamped at time $T_{4,k}$ in accordance with node A's clock. This process between the two nodes is repeated N times, where N stands for the required number of samples. It should be noted that N is a function of the target synchronization accuracy and the price the protocol is willing to invest in the form of network resources.

Based on the above pairwise synchronization message exchange mechanism, the clock offset measurement model can be represented in terms of the following two equations:

$$\begin{aligned} T_{2,k} &= T_{1,k} + d + \phi + X_k, \\ T_{4,k} &= T_{3,k} + d - \phi + Y_k. \end{aligned}$$

For simplicity, the above equations will be rewritten as

$$\begin{aligned} U_k &= d + \phi + X_k, \\ V_k &= d - \phi + Y_k, \end{aligned}$$

where the time differences U_k and V_k are defined as $U_k \triangleq T_{2,k} - T_{1,k}$ and $V_k \triangleq T_{4,k} - T_{3,k}$, respectively. The quantity d symbolizes the fixed portions of the delays assumed to be symmetric for each direction, X_k and Y_k denote the variable portions of delays and assume exponential distributions with means α and β, respectively, and ϕ stands for the clock offset of reference node B with respect to node A. During the interval between the pretransmission and postreception records of a timing cell, there are different kinds of incurred link delay uncertainties in the radio message delivery, which might assume magnitudes greater than the required precision of time synchronization. Therefore, it is very important to dig deeper into the exact nature and significance of all the components comprising these sources of error. Taking into account even the minutest details, Maroti *et al.* [58] classified all the link delay uncertainties incurred by the message as either deterministic or non-deterministic. The sources of delays such as send time, channel access time, interrupt handling time, receive time, etc., are non-deterministic and can range from around 5 μs to 500 ms. There are also deterministic sources of delays such as encoding time, transmission time, propagation time, reception time, decoding time, byte alignment time, etc., which can range from 0 μs to 20 ms. Besides Maroti *et al.*, numerous other authors, such as Abdel-Ghaffar [1] and Jeske [41] have divided the link delay uncertainties in deterministic and non-deterministic components. Interested readers are also encouraged to read [37] and [46] for a detailed study of network delays and their breakdown in detail.

Network delay modeling has been an active research topic since the 1980s. Out of the pdf models proposed to capture the distribution of the network delays, the Weibull, exponential, Gamma, and log-normal distributions ([13], [52], [70]) have received the greatest attention. There are various reasons behind choosing the exponential distribution for the purpose of this study. Moon *et al.* [64] collected several traces of delay measurements on the Internet and Mbone [26] for more than a month using constant length user datagram protocol (UDP) packets whose payloads consisted of a sequence number and a timestamp sent out at periodic intervals. The exponential distribution provided a quite satisfactory fit for the measurements obtained in the experiment. In addition, a single-server M/M/1 queue can appropriately model the cumulative link delay for point-to-point hypothetical reference connections, where the random delays are independently modeled as exponential random variables [1]. Moreover, Abdel-Ghaffar [1] proposed five different clock offset estimation algorithms, such as the median round delay, the minimum round delay, the minimum link delay, the median phase and the average phase, in which the minimum link delay algorithm has been experimentally demonstrated to be superior to the rest [71]. Jeske [41] later proved mathematically that the minimum link delay algorithm yields the ML estimate under exponential link delays. All these results confirm that the assumption of exponential distribution for network delays is an adequate model for a number of experimental observations and applications.

In [41], it was argued that for an unknown d, irrespective of the symmetric exponential distribution mean $\alpha = \beta \triangleq \lambda$ being known or unknown, the MLE of the vector parameter $\mathbf{\Phi}^S_{MLE} = [d\ \phi\ \lambda]$ is given by

$$\mathbf{\Phi}^S_{MLE} = \begin{bmatrix} \hat{d}^S_{MLE} \\ \hat{\phi}^S_{MLE} \\ \hat{\lambda}^S_{MLE} \end{bmatrix} = \frac{1}{2}\begin{bmatrix} U_{(1)} + V_{(1)} \\ U_{(1)} - V_{(1)} \\ \overline{U} + \overline{V} - \left(U_{(1)} + V_{(1)}\right) \end{bmatrix}, \tag{5.1}$$

where the superscript S represents the symmetric delay case, $U_{(1)}$ and $V_{(1)}$ denote the minimum order statistics, i.e., $U_{(1)} = \min_{1 \le i \le N} U_i$, $V_{(1)} = \min_{1 \le i \le N} V_i$, and $\overline{U} = (1/N)\sum_{k=1}^{N} U_i$ and $\overline{V} = (1/N)\sum_{k=1}^{N} V_i$ represent the sample average of the data $\{U_k\}_{k=1}^N$ and $\{V_k\}_{k=1}^N$, respectively. When λ is known, the MLE of $\{d, \phi\}$ remains the same.

Next, BLUE-OS and MVUE are derived for both asymmetric and symmetric cases, assuming known and unknown exponential delay means.

5.2 Best Linear Unbiased Estimation Using Order Statistics (BLUE-OS)

Deriving regular BLUE for a problem in general yields suboptimal results, since the class of unbiased estimators, within which the search is performed, is restricted to be linear. In the case when the noise is normally distributed, direct application of BLUE provides the optimal solution by virtue of the Gauss–Markov Theorem [44]. However, for other distributions, including the exponential distribution as is the case with the modeling framework adopted in this chapter, the application of BLUE might appear to lack any significance given the apparent loss of optimality in such a context. However, for a general location-scale distribution, Lloyd [55] suggested a new technique based on the derivation of BLUE using order statistics instead of just the raw observations. Such a technique will be applied herein to the target scenario as follows.

Let the order statistics of the observations $\{U_k\}_{k=1}^N$ and $\{V_k\}_{k=1}^N$ be denoted $\{U_{(k)}\}_{k=1}^N$ and $\{V_{(k)}\}_{k=1}^N$, respectively. Define

$$U'_k \triangleq \frac{1}{\alpha}(U_k - d - \phi), \tag{5.2}$$

$$V'_k \triangleq \frac{1}{\beta}(V_k - d + \phi), \tag{5.3}$$

which are a set of independent observations on the standardized variate and hence their distribution is parameter-free. The order statistics of U'_k and V'_k are denoted by $U'_{(k)}$ and $V'_{(k)}$, respectively. From (5.2) and (5.3), the following relations hold:

$$E\left[U_{(k)}\right] = d + \phi + \alpha E\left[U'_{(k)}\right], \quad E\left[V_{(k)}\right] = d - \phi + \beta E\left[V'_{(k)}\right],$$
$$\text{var}\left[U_{(k)}\right] = \alpha^2 \text{var}\left[\text{U}'_{(\text{k})}\right], \quad \text{var}\left[V_{(k)}\right] = \beta^2 \text{var}\left[\text{V}'_{(\text{k})}\right],$$
$$\text{cov}\left[U_{(k)} U_{(j)}\right] = \alpha^2 \text{cov}\left[U'_{(k)} U'_{(j)}\right], \quad \text{cov}\left[V_{(k)} V_{(j)}\right] = \beta^2 \text{cov}\left[V'_{(k)} V'_{(j)}\right].$$

Now using standard results from [42], the statistics of the ordered samples can be expressed as

$$\begin{aligned}
E\left[U'_{(k)}\right] &= E\left[V'_{(k)}\right] = \sum_{i=1}^{k} \frac{1}{(N-i+1)}, \\
\text{var}\left[U'_{(k)}\right] &= \text{var}\left[V'_{(k)}\right] = \sum_{i=1}^{k} \frac{1}{(N-i+1)^2}, \\
\text{cov}\left[U'_{(k)} U'_{(j)}\right] &= \text{cov}\left[V'_{(k)} V'_{(j)}\right] = \sum_{i=1}^{k} \frac{1}{(N-i+1)^2}.
\end{aligned}$$

As a result, the $N \times N$ symmetric positive-definite covariance matrix $\mathbf{C}$ for both $U'_{(k)}$ and $V'_{(k)}$, where $[C]_{kj} = \text{cov}[\text{U}'_{(\text{k})}\text{U}'_{(\text{j})}] = \text{cov}[\text{V}'_{(\text{k})}\text{V}'_{(\text{j})}]$, takes the form

$$\mathbf{C} = \begin{bmatrix}
\frac{1}{N^2} & \frac{1}{N^2} & \cdots & \frac{1}{N^2} \\
\frac{1}{N^2} & \frac{1}{N^2} + \frac{1}{(N-1)^2} & \cdots & \frac{1}{N^2} + \frac{1}{(N-1)^2} \\
\vdots & \vdots & \cdots & \vdots \\
\frac{1}{N^2} & \frac{1}{N^2} + \frac{1}{(N-1)^2} & \cdots & \sum_{k=1}^{N} \frac{1}{(N-k+1)^2}
\end{bmatrix}.$$

A simple exercise utilizing Gauss–Jordan elimination yields the following closed-form expression for the inverse of the covariance matrix:

$$\mathbf{C}^{-1} = \begin{bmatrix}
N^2 + (N-1)^2 & -(N-1)^2 & 0 & \cdots & 0 \\
-(N-1)^2 & (N-1)^2 + (N-2)^2 & -(N-2)^2 & \cdots & 0 \\
0 & -(N-2)^2 & (N-2)^2 + (N-3)^2 & \cdots & 0 \\
\vdots & \vdots & \vdots & \cdots & \vdots \\
0 & 0 & 0 & \cdots & 1
\end{bmatrix}.$$

The BLUE-OS will next be derived separately for both symmetric and asymmetric network delays.

5.2.1 Symmetric Link Delays

The symmetric network delay assumption holds true for some realistic scenarios, especially when the nodes have a direct communication link between them and the topology of the network is constant. In this case, $\alpha = \beta \triangleq \lambda$. Consider BLUE-OS $\mathbf{\Phi}^S_{BLUE-OS} \triangleq [d \; \phi \; \lambda]^T$, which is a linear function of an ordered set of observations $\{U_{(k)}\}_{k=1}^N$ and $\{V_{(k)}\}_{k=1}^N$. Let $\mathbf{z} \triangleq [U_{(1)} \; U_{(2)} \cdots U_{(N)} \; V_{(1)} \; V_{(2)} \cdots V_{(N)}]^T$. Then, it is straightforward to see that

$$E[\mathbf{z}] = \begin{bmatrix} 1 & 1 & \cdots & 1 & 1 & 1 & \cdots & 1 \\ 1 & 1 & \cdots & 1 & -1 & -1 & \cdots & -1 \\ \frac{1}{N} & \frac{1}{N} + \frac{1}{N-1} & \cdots & \sum_{k=1}^N \frac{1}{(N-k+1)} & \frac{1}{N} & \frac{1}{N} + \frac{1}{N-1} & \cdots & \sum_{k=1}^N \frac{1}{(N-k+1)} \end{bmatrix}^T \times \begin{bmatrix} d \\ \phi \\ \lambda \end{bmatrix} = \mathbf{Q}\mathbf{\Phi}^S_{BLUE-OS},$$

where $\mathbf{z}$ is the $2N \times 1$ ordered data vector, $\mathbf{Q}$ is a known matrix of dimension $2N \times 3$ and $\mathbf{\Phi}^S_{BLUE-OS}$ is the 3×1 vector of unknown parameters. The above linear relationship allows the problem to be solved using the Gauss–Markov theorem as follows:

$$\hat{\mathbf{\Phi}}^S_{BLUE-OS} = \left(\mathbf{Q}^T\mathbf{C_z}^{-1}\mathbf{Q}\right)^{-1}\mathbf{Q}^T\mathbf{C_z}^{-1}\mathbf{z}. \tag{5.4}$$

Since $\{U_{(k)}\}_{k=1}^N$ and $\{V_{(k)}\}_{k=1}^N$ are independent data sets, $\mathbf{C_z}$ is now given by

$$\mathbf{C_z} = \lambda^2 \begin{bmatrix} \mathbf{C} & \mathbf{0} \\ \mathbf{0} & \mathbf{C} \end{bmatrix},$$

and its inverse can be expressed as

$$\mathbf{C_z}^{-1} = \frac{1}{\lambda^2} \begin{bmatrix} \mathbf{C}^{-1} & \mathbf{0} \\ \mathbf{0} & \mathbf{C}^{-1} \end{bmatrix}.$$

It follows that

$$\mathbf{Q}^T\mathbf{C_z}^{-1}\mathbf{Q} = \frac{1}{\lambda^2} \begin{bmatrix} 2N^2 & 0 & 2N \\ 0 & 2N^2 & 0 \\ 2N & 0 & 2N \end{bmatrix},$$

and its inverse is

$$\left(\mathbf{Q}^T\mathbf{C_z}^{-1}\mathbf{Q}\right)^{-1} = \frac{\lambda^2}{2N(N-1)} \begin{bmatrix} 1 & 0 & -1 \\ 0 & \frac{N-1}{N} & 0 \\ -1 & 0 & N \end{bmatrix}. \tag{5.5}$$

This allows the multiplicative factor of $\mathbf{z}$ in (5.4) to be expressed in the form

$$\left(\mathbf{Q}^T\mathbf{C_z}^{-1}\mathbf{Q}\right)^{-1}\mathbf{Q}^T\mathbf{C_z}^{-1}$$
$$= \frac{1}{2N(N-1)}\begin{bmatrix} N^2-1 & -1 & \cdots & -1 & N^2-1 & -1 & \cdots & -1 \\ N(N-1) & 0 & \cdots & 0 & -N(N-1) & 0 & \cdots & 0 \\ N-N^2 & N & \cdots & N & N-N^2 & N & \cdots & N \end{bmatrix}.$$

Therefore, BLUE-OS in the symmetric exponential network delays case is given by

$$\begin{aligned} \hat{\mathbf{\Phi}}^S_{BLUE-OS} &= \begin{bmatrix} \hat{d}^S_{BLUE-OS} \\ \hat{\phi}^S_{BLUE-OS} \\ \hat{\lambda}^S_{BLUE-OS} \end{bmatrix} \\ &= \frac{1}{2N(N-1)}\begin{bmatrix} (N^2-1)\,U_{(1)} - \sum_{k=2}^{N} U_{(k)} + (N^2-1)\,V_{(1)} - \sum_{k=2}^{N} V_{(k)} \\ N(N-1)\,U_{(1)} - N(N-1)\,V_{(1)} \\ (N-N^2)\,U_{(1)} + \sum_{k=2}^{N} U_{(k)} + (N-N^2)\,V_{(1)} + \sum_{k=2}^{N} V_{(k)} \end{bmatrix} \\ &= \frac{1}{2(N-1)}\begin{bmatrix} N\left(U_{(1)}+V_{(1)}\right) - (\overline{U}+\overline{V}) \\ (N-1)\left(U_{(1)}-V_{(1)}\right) \\ N\left\{(\overline{U}+\overline{V}) - \left(U_{(1)}+V_{(1)}\right)\right\} \end{bmatrix}, \end{aligned} \tag{5.6}$$

where $\overline{U}$ and $\overline{V}$ represent the sample averages of the data sets $\{U_k\}_{k=1}^N$ and $\{V_k\}_{k=1}^N$, respectively, and coincide with the sample averages of ordered observations $\{U_{(k)}\}_{k=1}^N$ and $\{V_{(k)}\}_{k=1}^N$, respectively. Note that the derived BLUE-OS of the clock offset in (5.6) matches the expression of the MLE in (5.1).

5.2.2 Asymmetric Link Delays

In many broadband and wireless channels, and ad-hoc networks with time-varying topologies, the symmetric network delay assumption does not hold and applying the same results derived under the symmetric assumption is suboptimal. Therefore, a new method of deriving the efficient estimators in this case is of paramount importance. Let $\mathbf{\Phi}^A_{BLUE-OS} \triangleq [d\ \phi\ \alpha\ \beta]^T$ denote the vector of unknown parameters, then the linear model based on the ordered observations can be expressed as

$$E[\mathbf{z}] = \begin{bmatrix} 1 & 1 & \cdots & 1 & 1 & 1 & \cdots & 1 \\ 1 & 1 & \cdots & 1 & -1 & -1 & \cdots & -1 \\ \frac{1}{N} & \frac{1}{N}+\frac{1}{N-1} & \cdots & \sum_{k=1}^{N}\frac{1}{(N-k+1)} & 0 & 0 & \cdots & 0 \\ 0 & 0 & \cdots & 0 & \frac{1}{N} & \frac{1}{N}+\frac{1}{N-1} & \cdots & \sum_{k=1}^{N}\frac{1}{(N-k+1)} \end{bmatrix}^T \times \begin{bmatrix} d \\ \phi \\ \alpha \\ \beta \end{bmatrix} = \mathbf{Q}\mathbf{\Phi}_{BLUE-OS}^{A},$$

where $\mathbf{z}$ is again a $2N \times 1$ concatenated vector of ordered data $U_{(k)}$ and $V_{(k)}$, $\mathbf{z} \triangleq [U_{(1)}\ U_{(2)} \cdots U_{(N)}\ V_{(1)}\ V_{(2)} \cdots V_{(N)}]^T$, $\mathbf{Q}$ is a known matrix of dimension $2N \times 4$, and $\mathbf{\Phi}_{BLUE-OS}^{A}$ is the 4×1 vector of unknown parameters. Since the model has been shown to be linear in terms of the ordered observations, the BLUE-OS is now given by

$$\hat{\mathbf{\Phi}}_{BLUE-OS}^{A} = \left(\mathbf{Q}^T\mathbf{C_z}^{-1}\mathbf{Q}\right)^{-1}\mathbf{Q}^T\mathbf{C_z}^{-1}\mathbf{z}, \tag{5.7}$$

where $\mathbf{C_z}$ is the covariance matrix of vector $\mathbf{z}$. Due to the mutual independence of $U_{(k)}$ and $V_{(k)}$, $\mathbf{C_z}$ takes the form of the diagonal matrix

$$\mathbf{C_z} = \begin{bmatrix} \alpha^2\mathbf{C} & \mathbf{0} \\ \mathbf{0} & \beta^2\mathbf{C} \end{bmatrix},$$

and its inverse can be expressed as

$$\mathbf{C_z}^{-1} = \frac{1}{\alpha^2\beta^2}\begin{bmatrix} \beta^2\mathbf{C}^{-1} & \mathbf{0} \\ \mathbf{0} & \alpha^2\mathbf{C}^{-1} \end{bmatrix}. \tag{5.8}$$

Based on (5.8), it follows that

$$\mathbf{Q}^T\mathbf{C_z}^{-1}\mathbf{Q} = \begin{bmatrix} (\alpha^{-2}+\beta^{-2})N^2 & (\alpha^{-2}-\beta^{-2})N^2 & \alpha^{-2}N & \beta^{-2}N \\ (\alpha^{-2}-\beta^{-2})N^2 & (\alpha^{-2}+\beta^{-2})N^2 & \alpha^{-2}N & -\beta^{-2}N \\ \alpha^{-2}N & \alpha^{-2}N & \alpha^{-2}N & 0 \\ \beta^{-2}N & -\beta^{-2}N & 0 & \beta^{-2}N \end{bmatrix}$$

and its inverse takes the form

$$\left(\mathbf{Q}^T\mathbf{C_z}^{-1}\mathbf{Q}\right)^{-1} = \frac{1}{2N(N-1)}\begin{bmatrix} \frac{1}{2}(\alpha^2+\beta^2) & \frac{1}{2}(\alpha^2-\beta^2) & -\alpha^2 & -\beta^2 \\ \frac{1}{2}(\alpha^2-\beta^2) & \frac{1}{2}(\alpha^2+\beta^2) & -\alpha^2 & \beta^2 \\ -\alpha^2 & -\alpha^2 & 2N\alpha^2 & 0 \\ -\beta^2 & \beta^2 & 0 & 2N\beta^2 \end{bmatrix}. \tag{5.9}$$

Consequently, the multiplicative factor in the right hand side of (5.7) assumes the expression

$$\left(\mathbf{Q}^T\mathbf{C_z}^{-1}\mathbf{Q}\right)^{-1}\mathbf{Q}^T\mathbf{C_z}^{-1} = \frac{1}{2N(N-1)}\begin{bmatrix}\mathbf{A} & \mathbf{B}\end{bmatrix}.$$

where the matrices **A** and **B** are defined as

$$\mathbf{A} = \begin{bmatrix} \frac{N^2}{\alpha^2}\left(\frac{\alpha^2+\beta^2}{2}+\frac{\alpha^2-\beta^2}{2}\right)-1 & -1 & \cdots & -1 \\ \frac{N^2}{\alpha^2}\left(\frac{\alpha^2-\beta^2}{2}+\frac{\alpha^2+\beta^2}{2}\right)-1 & -1 & \cdots & -1 \\ 2N-2N^2 & 2N & \cdots & 2N \\ 0 & 0 & \cdots & 0 \end{bmatrix},$$

$$\mathbf{B} = \begin{bmatrix} \frac{N^2}{\beta^2}\left(\frac{\alpha^2+\beta^2}{2}-\frac{\alpha^2-\beta^2}{2}\right)-1 & -1 & \cdots & -1 \\ \frac{N^2}{\beta^2}\left(\frac{\alpha^2-\beta^2}{2}-\frac{\alpha^2+\beta^2}{2}\right)+1 & 1 & \cdots & 1 \\ 0 & 0 & \cdots & 0 \\ 2N-2N^2 & 2N & \cdots & 2N \end{bmatrix}.$$

It follows from the above equations that

$$\left(\mathbf{Q}^T\mathbf{C_z}^{-1}\mathbf{Q}\right)^{-1}\mathbf{Q}^T\mathbf{C_z}^{-1}$$
$$= \frac{1}{2N(N-1)}\begin{bmatrix} N^2-1 & -1 & \cdots & -1 & N^2-1 & -1 & \cdots & -1 \\ N^2-1 & -1 & \cdots & -1 & -(N^2-1) & 1 & \cdots & 1 \\ 2N-2N^2 & 2N & \cdots & 2N & 0 & 0 & \cdots & 0 \\ 0 & 0 & \cdots & 0 & 2N-2N^2 & 2N & \cdots & 2N \end{bmatrix},$$

which according to (5.7) leads to

$$\hat{\mathbf{\Phi}}^A_{BLUE-OS} = \begin{bmatrix} \hat{d}^A_{BLUE-OS} \\ \hat{\phi}^A_{BLUE-OS} \\ \hat{\alpha}^A_{BLUE-OS} \\ \hat{\beta}^A_{BLUE-OS} \end{bmatrix}$$
$$= \frac{1}{2N(N-1)}\begin{bmatrix} (N^2-1)\,U_{(1)} - \sum\limits_{k=2}^{N} U_{(k)} + (N^2-1)\,V_{(1)} - \sum\limits_{k=2}^{N} V_{(k)} \\ (N^2-1)\,U_{(1)} - \sum\limits_{k=2}^{N} U_{(k)} - (N^2-1)\,V_{(1)} + \sum\limits_{k=2}^{N} V_{(k)} \\ 2NU_{(1)} - 2N^2U_{(1)} + 2N\sum\limits_{k=2}^{N} U_{(k)} \\ 2NV_{(1)} - 2N^2V_{(1)} + 2N\sum\limits_{k=2}^{N} V_{(k)} \end{bmatrix}$$

$$= \frac{1}{2(N-1)} \begin{bmatrix} N\left(U_{(1)} + V_{(1)}\right) - (\overline{U} + \overline{V}) \\ N\left(U_{(1)} - V_{(1)}\right) - (\overline{U} - \overline{V}) \\ 2N\left(\overline{U} - U_{(1)}\right) \\ 2N\left(\overline{V} - V_{(1)}\right) \end{bmatrix}. \tag{5.10}$$

5.3 Minimum Variance Unbiased Estimation (MVUE)

In parameter estimation, very often the ultimate goal is to find the estimator that achieves the minimum MSE. However, it is well known in theory that the optimal MSE estimators are usually not realizable. Since the MSE is the sum of the estimator variance and the squared bias, a technique chosen to attain realizable yet best estimators is to constrain the bias to be zero (since the dependence of the minimum MSE estimator on the unknown parameter typically comes from the bias). Therefore, restricting the possible estimators to be unbiased and then finding the estimator with the smallest variance for all values of the unknown parameter yields the optimal solution within the class of unbiased estimators. Therefore, we will focus on MVUE.

The "turn-the-crank" procedure for deriving the MVUE in estimation theory is based on the Rao–Blackwell–Lehmann–Scheffé theorem. First, the likelihood function is factored according to Neymann–Fisher factorization theorem yielding the sufficient statistics $\mathbf{T}$. Then, it is determined whether the sufficient statistics are complete. Finally, either for any unbiased estimator $\check{\theta}$, $\hat{\theta} = E[\check{\theta}|\mathbf{T}]$ is evaluated, or a function $g(\mathbf{T})$ of the sufficient statistics is found such that $\hat{\theta} = g(\mathbf{T})$ is an unbiased estimator, producing $\hat{\theta}$ as the MVUE. The approach that we will follow next relies on similar steps.

5.3.1 Asymmetric Link Delays

Starting with the asymmetric case, the likelihood function for the clock offset as a function of observations $\{U_k\}_{k=1}^N$ and $\{V_k\}_{k=1}^N$ is given by

$$\begin{aligned} L(d,\phi,\alpha,\beta) &= \alpha^{-N}\exp\left[-\frac{1}{\alpha}\sum_{k=1}^{N}\{U_k - d - \phi\}\right]\beta^{-N}\exp\left[-\frac{1}{\beta}\sum_{k=1}^{N}\{V_k - d + \phi\}\right] \\ &\quad \times I\left[U_{(1)} - d - \phi\right] I\left[V_{(1)} - d + \phi\right], \end{aligned} \tag{5.11}$$

where $I[\cdot]$ denotes the unit step function. Exploiting the fact that the raw sample mean and the ordered sample mean are actually the same, (5.11) can be factored as

$$L(d,\phi,\alpha,\beta) = g_1\left(\sum_{k=1}^{N} U_{(k)}, d, \phi, \alpha\right) g_2\left(\sum_{k=1}^{N} V_{(k)}, d, \phi, \beta\right) g_3(U_{(1)}, d, \phi) g_4(V_{(1)}, d, \phi) \times h_1(U_k, V_k),$$

where

$$g_1\left(\sum_{k=1}^{N} U_{(k)}, d, \phi, \alpha\right) = \alpha^{-N} \exp\left[-\frac{1}{\alpha}\sum_{k=1}^{N}\left(U_{(k)} - d - \phi\right)\right], g_3\left(U_{(1)}, d, \phi\right) = I\left[U_{(1)} - d - \phi\right],$$

$$g_2\left(\sum_{k=1}^{N} V_{(k)}, d, \phi, \beta\right) = \beta^{-N} \exp\left[-\frac{1}{\beta}\sum_{k=1}^{N}\left(V_{(k)} - d + \phi\right)\right], g_4\left(V_{(1)}, d, \phi\right) = I\left[V_{(1)} - d + \phi\right],$$

$$h_1\left(U_k, V_k\right) = 1.$$

In the above relations, the factor $h_1(U_k, V_k)$ is independent of the unknown vector parameter $\mathbf{\Phi}_{MVUE}^{A-U} = [d\ \phi\ \alpha\ \beta]^T$, whereas $g_1(\sum_{k=1}^{N} U_{(k)}, d, \phi, \alpha)$, $g_2(\sum_{k=1}^{N} V_{(k)}, d, \phi, \beta)$, $g_3(U_{(1)}, d, \phi)$ and $g_4(V_{(1)}, d, \phi)$ are functions depending on the data through $\mathbf{T} = \{\sum_{k=1}^{N} U_{(k)}, U_{(1)}, \sum_{k=1}^{N} V_{(k)}, V_{(1)}\}$. Therefore, according to Neymann–Fisher factorization theorem, $\mathbf{T}$ is a sufficient statistic for $\mathbf{\Phi}_{MVUE}^{A-U}$.

Since $\dim(\mathbf{T}) = \dim(\mathbf{\Phi}_{MVUE}^{A-U})$, it is easier to determine the MVUE directly from $\mathbf{T}$ without having to evaluate $E[\check{\mathbf{\Phi}}_{MVUE}^{A-U}|\mathbf{T}]$ by finding a 4×1 vector function $\hat{\mathbf{\Phi}}_{MVUE}^{A-U}$ such that $E[\hat{\mathbf{\Phi}}_{MVUE}^{A-U}] = \mathbf{\Phi}_{MVUE}^{A-U}$, provided that $\mathbf{T}$ is a complete sufficient statistic. Finding the pdf of $\mathbf{T}$ is required to prove that $\mathbf{T}$ is complete, but the problem of finding this pdf is a little complex, because $\sum_{k=1}^{N} U_{(k)}$ and $U_{(1)}$, and similarly $\sum_{k=1}^{N} V_{(k)}$ and $V_{(1)}$, are not independent.

The joint pdf of $U_{(1)}, U_{(2)}, \ldots, U_{(N)}$ is given by

$$p\left(U_{(1)}, U_{(2)}, \ldots, U_{(N)}\right) = N!\alpha^{-N} \exp\left[-\frac{1}{\alpha}\sum_{k=1}^{N}\left\{U_{(k)} - d - \phi\right\}\right] \prod_{k=1}^{N} I\left[U_{(k)} - d - \phi\right], \tag{5.12}$$

whereas the pdf of the minimum order statistic $U_{(1)}$ is also exponential with mean α/N. Now consider the transformation

$$z_k = (N-k+1)\left(U_{(k)} - U_{(k-1)}\right), \quad k = 1, 2, \ldots, N, \tag{5.13}$$

where $U_{(0)} = d + \phi$. Since $\sum_{k=1}^{N}(U_{(k)} - d - \phi) = \sum_{k=1}^{N} z_k$ and the Jacobian of the transformation is $N!$, a substitution of (5.13) in (5.12) reveals that

$$p\left(z_1, z_2, \ldots, z_N\right) = \alpha^{-N} \exp\left(-\frac{1}{\alpha}\sum_{k=1}^{N} z_k\right) \prod_{k=1}^{N} I\left[z_k\right],$$

i.e., z_k are *independent* exponential random variables with similar mean α. In addition, since each $z_k \sim \exp(\alpha)$, each z_k assumes a Gamma distribution $z_k \sim \Gamma(1, \alpha)$, too. Using the relationship $\sum_{k=1}^{N}(U_{(k)} - U_{(1)}) = \sum_{k=2}^{N} z_k$, and the fact that each of $z_2, z_3, \ldots, z_N$ is independent of z_1 (and hence of $U_{(1)}$, since $z_1 = N(U_{(1)} - d - \phi)$), $\sum_{k=1}^{N}(U_{(k)} - U_{(1)}) \sim \Gamma(N-1, \alpha)$ and is independent of $U_{(1)}$.

By a similar reasoning, it can be inferred that $\sum_{k=1}^{N}(V_{(k)} - V_{(1)}) \sim \Gamma(N-1, \beta)$ and is independent of $V_{(1)}$. Therefore, the one-to-one function $\mathbf{T}' = \{\sum_{k=1}^{N}(U_{(k)} - U_{(1)}), U_{(1)}, \sum_{k=1}^{N}(V_{(k)} - V_{(1)}), V_{(1)}\}$ of $\mathbf{T}$ is also sufficient for estimating $\mathbf{\Phi}_{MVUE}^{A-U}$ because the sufficient statistics is unique within one-to-one transformations [44]. Consequently, $\mathbf{T}'$ comprises four independent random variables that in terms of the three-parameter Gamma distribution assume the distributions:

$$r = \sum_{k=1}^{N}(U_{(k)} - U_{(1)}) \sim \Gamma\left(N-1, \alpha, 0\right), \qquad s = \sum_{k=1}^{N}(V_{(k)} - V_{(1)}) \sim \Gamma\left(N-1, \beta, 0\right),$$
$$U_{(1)} \sim \Gamma\left(1, \alpha/N, d+\phi\right), \qquad V_{(1)} \sim \Gamma\left(1, \beta/N, d-\phi\right).$$

Note that the domains of r and s are controlled by $U_{(1)}$ and $V_{(1)}$, respectively. Next, whether $\mathbf{T}'$, or equivalently $\mathbf{T}$, is complete has to be checked. Completeness implies that there is but one function of $\mathbf{T}$ that is unbiased. Let $g(\mathbf{T}')$ be a function of $\mathbf{T}'$ such that $E[g(\mathbf{T}')] = \mathbf{\Phi}_{MVUE}^{A-U}$. Suppose that there exists another function h for which $E[h(\mathbf{T}')] = \mathbf{\Phi}_{MVUE}^{A-U}$ is also true. Then,

$$E\left[g\left(\mathbf{T}'\right) - h\left(\mathbf{T}'\right)\right] = E\left[\pi\left(\mathbf{T}'\right)\right] = 0, \qquad \forall\ \mathbf{\Phi}_{MVUE}^{A-U},$$

where $\pi(\mathbf{T}') \triangleq g(\mathbf{T}') - h(\mathbf{T}')$ and the expectation is taken with respect to $p(\mathbf{T}'; \mathbf{\Phi}_{MVUE}^{A-U})$. As a result,

$$\begin{aligned}
&\underset{R_{U_{(1)},V_{(1)}}}{\int\int\int\int} \pi\left(r, U_{(1)}, s, V_{(1)}\right) \frac{\alpha^{-(N-1)}}{\Gamma\left(N-1\right)} r^{N-2} \exp\left(-\frac{r}{\alpha}\right) \frac{N}{\alpha} \\
&\times \exp\left(-\frac{N}{\alpha}\left\{U_{(1)} - d - \phi\right\}\right) \frac{\beta^{-(N-1)}}{\Gamma\left(N-1\right)} s^{N-2} \exp\left(-\frac{s}{\beta}\right) \frac{N}{\beta} \\
&\times \exp\left(-\frac{N}{\beta}\left\{V_{(1)} - d + \phi\right\}\right) dr\, dU_{(1)}\, ds\, dV_{(1)} \;=\; 0\ \forall\ \mathbf{\Phi}_{MVUE}^{A-U},
\end{aligned}$$

where $R_{U_{(1)},V_{(1)}}$ is the region defined by $I[U_{(1)} - d - \phi]$ and $I[V_{(1)} - d - \phi]$. The above relation can be expressed as

$$\int_{-\infty}^{\infty}\int_{-\infty}^{\infty}\int_{-\infty}^{\infty}\int_{-\infty}^{\infty} \pi\left(r, U_{(1)}, s, V_{(1)}\right) r^{N-2} s^{N-2} \exp\left[-\left\{\frac{r}{\alpha} + \frac{NU_{(1)}}{\alpha} + \frac{s}{\beta} + \frac{NV_{(1)}}{\beta}\right\}\right]$$
$$\times\, dr\, dU_{(1)}\, ds\, dV_{(1)} = 0, \qquad \forall\ \mathbf{\Phi}^{A-U}_{MVUE}. \tag{5.14}$$

The expression on the left hand side of (5.14) is the four-dimensional Laplace transform of the function $\pi(\mathbf{T}')$. It follows from the uniqueness theorem for a two-sided Laplace transform that $\pi(\mathbf{T}') = 0$ almost everywhere, resulting in $g(\mathbf{T}') = h(\mathbf{T}')$ and hence there is only one unbiased function of $\mathbf{T}'$. This proves that the statistic $\mathbf{T}'$, or equivalently $\mathbf{T}$, is complete for estimating $\mathbf{\Phi}^{A-U}_{MVUE}$ when the links are asymmetric and both α and β are unknown.

Finally, the complete sufficient statistic $\mathbf{T}$ is also minimal owing to Bahadur's theorem [9], which states that *if* $\mathbf{T}$, *taking values in* $\Re^k$, *is sufficient for* $\mathbf{\Phi}^{A-U}_{MVUE}$ *and boundedly complete, then* $\mathbf{T}$ *is minimal sufficient*.

What remains is finding an unbiased estimator for $\mathbf{\Phi}^{A-U}_{MVUE}$ as a function of $\mathbf{T}$, which is the MVUE according to the Rao–Blackwell–Lehmann–Scheffé theorem. At first, it may seem difficult to find four unbiased functions of $\mathbf{T}$ for each of d, ϕ, α, and β just by inspection. But note that BLUE-OS $\hat{\mathbf{\Phi}}^{A}_{BLUE-OS}$ in (5.10) is also an unbiased function of $\mathbf{T}$. Hence, it is concluded that the BLUE-OS is also the MVUE:

$$\hat{\mathbf{\Phi}}^{A-U}_{MVUE} = \begin{bmatrix} \hat{d}^{A-U}_{MVUE} \\ \hat{\phi}^{A-U}_{MVUE} \\ \hat{\alpha}^{A-U}_{MVUE} \\ \hat{\beta}^{A-U}_{MVUE} \end{bmatrix} = \frac{1}{2(N-1)} \begin{bmatrix} N\left(U_{(1)} + V_{(1)}\right) - (\overline{U} + \overline{V}) \\ N\left(U_{(1)} - V_{(1)}\right) - (\overline{U} - \overline{V}) \\ 2N\left(\overline{U} - U_{(1)}\right) \\ 2N\left(\overline{V} - V_{(1)}\right) \end{bmatrix}. \tag{5.15}$$

The covariance matrix of this estimator is given by (5.9) and hence minimum variances of the clock offsets, fixed and mean delay parameters are given by its diagonal elements, whereas the total mean square error for the vector parameter $\hat{\mathbf{\Phi}}^{A-U}_{MVUE}$ is the trace of this matrix.

As a result, the MVUE for the desired parameter, the clock offset, for asymmetric unknown network delays is expressed as

$$\hat{\phi}^{A-U}_{MVUE} = \frac{1}{N-1}\left[N\frac{U_{(1)} - V_{(1)}}{2} - \frac{\overline{U} - \overline{V}}{2}\right], \tag{5.16}$$

and its variance or MSE is written as

$$\mathrm{var}\left(\hat{\phi}^{A-U}_{MVUE}\right) = \frac{1}{4N(N-1)}\left(\alpha^2 + \beta^2\right).$$

Similarly, the MVUE of the fixed delay d and mean link delays α and β are the same as in (5.10). For the sake of completeness, the MVUE is also given when α and β are known. It is straightforward to see from (5.11) that $U_{(1)}$ and $V_{(1)}$ are the complete minimal sufficient statistic for estimating d and ϕ. The only unbiased functions of $\{U_{(1)}, V_{(1)}\}$ yielding $\mathbf{\Phi}^{A-K}_{MVUE}$ are

$$\hat{\mathbf{\Phi}}^{A-K}_{MVUE} = \begin{bmatrix} \hat{d}^{A-K}_{MVUE} \\ \hat{\phi}^{A-K}_{MVUE} \end{bmatrix} = \frac{1}{2} \begin{bmatrix} \left(U_{(1)} - \frac{\alpha}{N}\right) + \left(V_{(1)} - \frac{\beta}{N}\right) \\ \left(U_{(1)} - \frac{\alpha}{N}\right) - \left(V_{(1)} - \frac{\beta}{N}\right) \end{bmatrix}. \tag{5.17}$$

5.3.2 SYMMETRIC LINK DELAYS

In the symmetric case when $\alpha = \beta \triangleq \lambda$, the likelihood function for the clock offset as a function of observations $\{U_k\}_{k=1}^N$ and $\{V_k\}_{k=1}^N$ is given by

$$L(d,\phi,\lambda) = \lambda^{-2N} \exp\left[-\frac{1}{\lambda}\sum_{k=1}^{N}\{U_k + V_k - 2d\}\right] I\left[U_{(1)} - d - \phi\right] I\left[V_{(1)} - d + \phi\right]. \tag{5.18}$$

For unknown λ, it seems that $\{\sum_{k=1}^N U_k, U_{(1)}, \sum_{k=1}^N V_k, V_{(1)}\}$ are again the sufficient statistics for the estimation of $\mathbf{\Phi}^{S-U}_{MVUE} = [d\ \phi\ \lambda]^T$, but then they have already generated *two* unbiased clock offset estimators, given by (5.6) and (5.10). Naturally, the question arises: since the same sufficient statistics have been proved complete, how can they yield two unbiased estimators? The answer to this lies in the consistency of science when we note that (5.18) can be factored as

$$L(d,\phi,\alpha,\beta) = g_1\left(\sum_{k=1}^{N} U_{(k)}, \sum_{k=1}^{N} V_{(k)}, d, \lambda\right) g_2\left(U_{(1)}, d, \phi\right) g_3\left(V_{(1)}, d, \phi\right) h_1\left(U_k, V_k\right),$$

where

$$g_1\left(\sum_{k=1}^{N} U_{(k)}, \sum_{k=1}^{N} V_{(k)}, d, \lambda\right) = \lambda^{-2N} \exp\left[-\frac{1}{\lambda}\sum_{k=1}^{N}\{U_k + V_k - 2d\}\right],$$

$$g_2\left(U_{(1)}, d, \phi\right) = I\left[U_{(1)} - d - \phi\right],\ g_3\left(V_{(1)}, d, \phi\right) = I\left[V_{(1)} - d + \phi\right],\ h_1\left(U_k, V_k\right) = 1.$$

It turns out that $\mathbf{T} = \{\sum_{k=1}^N (U_k + V_k), U_{(1)}, V_{(1)}\}$ is the actual minimal sufficient statistic instead of $\{\sum_{k=1}^N U_k, U_{(1)}, \sum_{k=1}^N V_k, V_{(1)}\}$ according to Neymann–Fisher factorization theorem. Consequently, the clock offset estimator in (5.10) is not even a choice to consider for not being function of $\mathbf{T}$.

Now proceeding similarly to before, $\sum_{k=1}^{N}(U_k + V_k)$ is dependent on both $U_{(1)}$ and $V_{(1)}$. As a result, $\mathbf{T}$ can be transformed into $\mathbf{T}' = \{\sum_{k=1}^{N}(U_k - U_{(1)} + V_k - V_{(1)}), U_{(1)}, V_{(1)}\}$. It is evident from the reasoning in the last subsection that $\sum_{k=1}^{N}(U_k - U_{(1)} + V_k - V_{(1)})$ is Gamma distributed with parameters $2(N-1)$ and λ. Hence, $\mathbf{T}'$ is a combination of three independent RVs, which in terms of the three-parameter Gamma distribution assume the distributions

$$r = \sum_{k=1}^{N}(U_k - U_{(1)} + V_k - V_{(1)}) \sim \Gamma\left(2\left(N-1\right), \lambda, 0\right),$$
$$U_{(1)} \sim \Gamma\left(1, \lambda/N, d+\phi\right), \quad V_{(1)} \sim \Gamma\left(1, \lambda/N, d-\phi\right).$$

Next, defining $g(\mathbf{T}')$ and $h(\mathbf{T}')$ as functions of $\mathbf{T}'$ such that $E[g(\mathbf{T}')] = E[h(\mathbf{T}')] = \mathbf{\Phi}_{MVUE}^{S-U}$,

$$E\left[g\left(\mathbf{T}'\right) - h\left(\mathbf{T}'\right)\right] = E\left[\pi\left(\mathbf{T}'\right)\right] = 0 \qquad \forall\ \mathbf{\Phi}_{MVUE}^{S-U},$$

where the expectation is taken with respect to $p_{\mathbf{T}'}(\mathbf{T}'; \mathbf{\Phi}_{MVUE}^{S-U})$. As a result, since the domains of r and s are controlled by $U_{(1)}$ and $V_{(1)}$, respectively, it turns out that

$$\int\int_{R_{U_{(1)},V_{(1)}}}\int \pi\left(r, U_{(1)}, V_{(1)}\right) \frac{\lambda^{-\{2(N-1)\}}}{\Gamma\left[2\left(N-1\right)\right]} r^{2N-3} \exp\left(-\frac{r}{\lambda}\right) \left(\frac{N}{\lambda}\right)^2$$
$$\times \exp\left(-\frac{N}{\lambda}\left\{U_{(1)} + V_{(1)} - 2d\right\}\right) dr\, dU_{(1)}\, dV_{(1)} \quad = \quad 0 \quad \forall\ \mathbf{\Phi}_{MVUE}^{S-U},$$

where $R_{U_{(1)},V_{(1)}}$ is the region defined by $I[U_{(1)} - d - \phi]$ and $I[V_{(1)} - d - \phi]$. It follows that

$$\int_{-\infty}^{\infty}\int_{-\infty}^{\infty}\int_{-\infty}^{\infty} \pi\left(r, U_{(1)}, V_{(1)}\right) r^{2N-3} \exp\left(-\frac{N}{\lambda}\left\{\frac{r}{N} + U_{(1)} + V_{(1)}\right\}\right)$$
$$\times\, dr\, dU_{(1)}\, dV_{(1)} \;=\; 0,\ \forall\ \mathbf{\Phi}_{MVUE}^{S-U}.$$

From the uniqueness theorem for the two-sided Laplace transform, it follows that $\pi(\mathbf{T}') = 0$ almost everywhere, resulting in the completeness of $\mathbf{T}'$, or equivalently $\mathbf{T}$. Hence, $\mathbf{T}$ is also the minimal sufficient statistics from Bahadur's theorem and the MVUE is the same as $\hat{\mathbf{\Phi}}_{BLUE-OS}^{S}$ in (5.6) expressed as

$$\hat{\mathbf{\Phi}}_{MVUE}^{S-U} = \begin{bmatrix} \hat{d}_{MVUE}^{S-U} \\ \hat{\phi}_{MVUE}^{S-U} \\ \hat{\lambda}_{MVUE}^{S-U} \end{bmatrix} = \frac{1}{2\left(N-1\right)} \begin{bmatrix} N\left(U_{(1)} + V_{(1)}\right) - \left(\overline{U} + \overline{V}\right) \\ \left(N-1\right)\left(U_{(1)} - V_{(1)}\right) \\ N\left\{\left(\overline{U} + \overline{V}\right) - \left(U_{(1)} + V_{(1)}\right)\right\} \end{bmatrix}. \tag{5.19}$$

The covariance matrix of this estimator is given by (5.5) and the diagonal elements represent the variance of each unknown parameter, whereas the trace of this matrix is the total mean square error or variance for the vector parameter $\mathbf{\Phi}_{MVUE}^{S-U}$.

Hence, the MVUE for the clock offset, in the case of symmetric unknown network delays, is expressed as

$$\hat{\phi}_{MVUE}^{S-U} = \frac{U_{(1)} - V_{(1)}}{2}, \tag{5.20}$$

and its variance or MSE is given by

$$\mathrm{var}\left(\hat{\phi}_{MVUE}^{S-U}\right) = \frac{\lambda^2}{2N^2}.$$

Furthermore, the MVUEs for the fixed delay d and mean link delay λ under the symmetric assumption match the ones in (5.6). Finally, following a similar procedure, when λ is known, the sufficient statistics are $U_{(1)}$ and $V_{(1)}$ and the MVUE is

$$\hat{\mathbf{\Phi}}_{MVUE}^{S-K} = \begin{bmatrix} \hat{d}_{MVUE}^{S-K} \\ \hat{\phi}_{MVUE}^{S-K} \end{bmatrix} = \frac{1}{2}\begin{bmatrix} U_{(1)} + V_{(1)} \\ U_{(1)} - V_{(1)} \end{bmatrix}, \tag{5.21}$$

where the superscript K represents the case of known λ.

5.4 Explanatory Remarks

Summarizing the results derived so far, Tables 5.1 and 5.2 show the MVUE for the clock offset for the possible combinations of symmetries/asymmetries in the network delays and knowledge of the mean link delay parameters from equations (5.15), (5.17), (5.19), and (5.21).

It is evident from Tables 5.1 and 5.2 that in practical scenarios where the means of the exponentially distributed delays are unknown, the MVUE is given by (5.16) or (5.20) depending on whether the network delays are asymmetric or symmetric. The natural question arises at this stage: which estimator is better when these network delays are *slightly* asymmetric? To answer this question, note that the MVUE is not always the best estimator, it is only the best of the unbiased estimators. If some biased estimator is devised with reduced variance relative to the MVUE at the price of an insignificant increase in the squared bias, then the biased estimator might outperform the MVUE in the MSE sense. Hence, for the *asymmetric* unknown mean link delays case, we will compare the MSE of the MLE in (5.1) with the MVUE in (5.16) as follows:

$$MSE\left(\hat{\phi}_{MVUE}^{A-U}\right) = \frac{1}{4N(N-1)}\left(\alpha^2+\beta^2\right), \tag{5.22}$$

$$\begin{aligned} MSE\left(\hat{\phi}_{MLE}\right) &= \frac{1}{4N^2}\left(\alpha^2+\beta^2\right) + \frac{1}{4N^2}\left(\alpha-\beta\right)^2 \\ &= \frac{1}{2N^2}\left(\alpha^2+\beta^2-\alpha\beta\right). \end{aligned} \tag{5.23}$$

Table 5.1: *The MVUE for the clock offset for known mean link delay*

Clock offset		Delay mean known
Symmetric delays	MVUE	$\left(U_{(1)} - V_{(1)}\right)/2$
	MSE	$\lambda^2/2N^2$
	Remarks	Same as MLE
Asymmetric delays	MVUE	$\left[\left(U_{(1)} - \frac{\alpha}{N}\right) - \left(V_{(1)} - \frac{\beta}{N}\right)\right]/2$
	MSE	$(\alpha^2 + \beta^2)/4N^2$
	Remarks	Bias-compensated MLE

Table 5.2: *The MVUE for the clock offset for unknown mean link delay*

Clock offset		Delay mean unknown
Symmetric delays	MVUE	$\left(U_{(1)} - V_{(1)}\right)/2$
	MSE	$\lambda^2/2N^2$
	Remarks	Same as MLE and BLUE-OS
Asymmetric delays	MVUE	$\left[N\left(U_{(1)} - V_{(1)}\right) - (\overline{U} - \overline{V})\right]/2(N-1)$
	MSE	$(\alpha^2 + \beta^2)/4N(N-1)$
	Remarks	Same as BLUE-OS

Notice that though $\hat{\phi}_{MLE}$ is biased in the most realistic setting, i.e., in the case of asymmetric unknown mean link delays, in accordance with (5.22) and (5.23), it outperforms the MVUE under the condition

$$MSE\left(\hat{\phi}^{A-U}_{MVUE}\right) > MSE\left(\hat{\phi}_{MLE}\right),$$
$$\frac{1}{4N(N-1)}\left(\alpha^2 + \beta^2\right) > \frac{1}{2N^2}\left(\alpha^2 + \beta^2 - \alpha\beta\right),$$

which can be expressed equivalently as:

$$\frac{N}{2} - 1 < \frac{\alpha\beta}{(\alpha - \beta)^2} \triangleq f(\alpha, \beta). \tag{5.24}$$

The above relations give rise to a number of remarks. First, (5.24) provides the number of timing synchronization messages N to be exchanged given α and β, up to which the MLE has a smaller MSE than the MVUE for asymmetric link delays. It

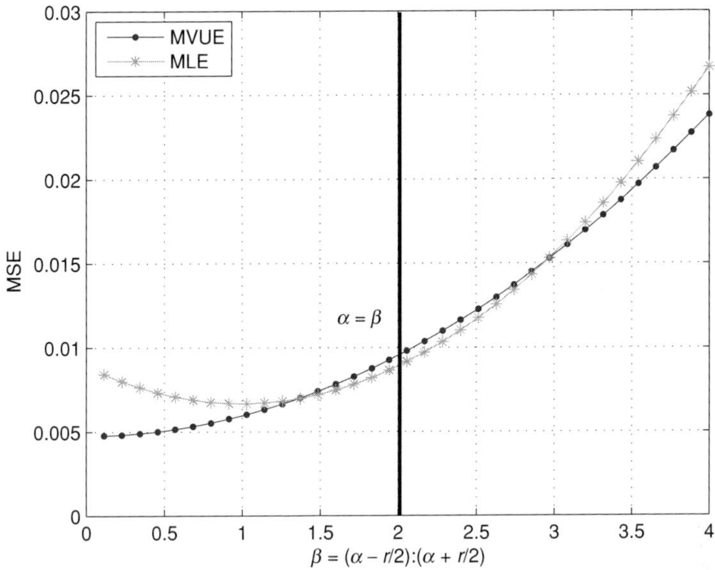

Figure 5.2: Mean square error (MSE) of the MVUE (5.22) and the MLE (5.23) for asymmetric unknown delays with constant $N = 15$.

also suggests that though the MLE is equal to the MVUE *only* in the symmetric link delays case, it attains a smaller MSE in the asymmetric case in the region around the point $\alpha = \beta$. As the asymmetry of the link increases, i.e., $|\alpha - \beta|$ tends to drift away from zero, the MVUE starts outperforming the MLE. The exact point at which their performances are the same can easily be derived from (5.24). The two respective MSEs are drawn in Figure 5.2, where N and α are held constant at 15 and 2, respectively, while β is varied across α through the relation $\beta \in [\alpha - r/2, \alpha + r/2]$. For this plot, the range r is chosen to be 4 and the step size is $r/70$. Figure 5.2 shows that the MSE of MLE actually initially decreases when β approaches α because the chosen $\alpha = 2$ is a small value and hence the MSE rise due to a slight increase in β is overcome by the fall in the MSE due to the smaller $|\alpha - \beta|$ (for larger values of α, this fall does not occur). It is clear that around the region where $\alpha = \beta$ (illustrated by the bold line in Figure 5.2), the MLE outperforms the MVUE and then a further increase in β again results in higher asymmetry thus making the MVUE the better choice. Second, it is evident from (5.22) and (5.23) that for a constant N, and increasing α and/or β, the MLE again exhibits better performance than the MVUE, and hence should be preferred over MVUE in networks with large delays. Third, (5.24) shows that for any $\alpha \neq \beta$, N can be made large enough to exceed the expression on the right hand side. This fact is also clear from Figure 5.3, where the same plot is drawn with N ranging from 15 to 20. Notice that although the MSEs of both estimators decrease with N, the two lines representing the intersections of the MSE curves

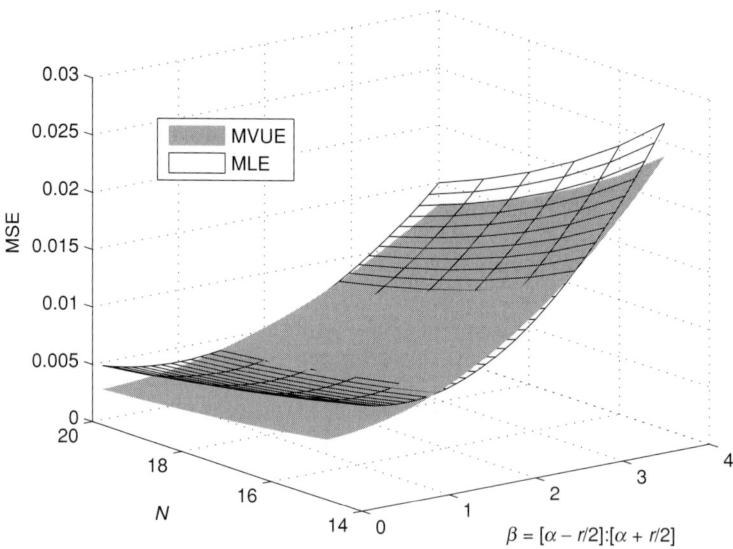

Figure 5.3: Mean square error (MSE) of the MLE (5.1) and the MVUE (5.16) for asymmetric unknown delays with different values of N.

manifest decreasing separation between them. This result corroborates the fact that MVUE overtakes the MLE after a certain number of observations. Fourth, it seems that for a constant N, estimating α and β utilizing (5.15) and (5.19) and plugging them into (5.24) might be a good idea for adaptively selecting between the MVUE and MLE as follows:

$$\frac{N}{2} - 1 \underset{MLE}{\overset{MVUE}{\gtrless}} \frac{\hat{\alpha}^{A-U}_{MVUE}\hat{\beta}^{A-U}_{MVUE}}{\left(\hat{\alpha}^{A-U}_{MVUE} - \hat{\beta}^{A-U}_{MVUE}\right)^2} = f(\hat{\alpha}^{A-U}_{MVUE}, \hat{\beta}^{A-U}_{MVUE}).$$

However, since $f(\hat{\alpha}, \hat{\beta})$ processes the estimates non-linearly, a considerable amplification of estimation errors occurs which affects the quality of the resultant $f(\hat{\alpha}, \hat{\beta})$. In other words, even having access to $\hat{\alpha}_{MLE}$ and $\hat{\beta}_{MLE}$ does not help to estimate $\hat{f}(\alpha, \beta)$ accurately, despite the fact that the MLE is functionally invariant. Nevertheless, such a technique can be used when the asymmetry between the delays is large, since the incorrect choice appears only around the region where the two MSE curves (as in Figure 5.2) intersect. These findings are very important in the context of WSNs where energy resources are limited and the number of synchronization packet exchanges is rather small. Even in the traditional centralized or ad-hoc networks, it should be noted that for an α fairly close to β, the MLE gives better results no matter by how much the magnitude of N is increased. In addition, when the topology of the network does not remain constant for long periods of time, as in ad-hoc networks, different delay environments are present during different synchronization cycles and choosing between the MVUE and the MLE according to each situation yields a better solution.

Based on the above observations, it should be emphasized that the problem under study provides an excellent textbook example of the worth of the MLE in real-world scenarios. It is not only relatively easy to derive, but it also performs outstandingly well in comparison to other laboriously obtained optimal (in some sense) estimators. This is why it has been the most widely used estimator to date in engineering applications.

As a final remark, note that our primary interest was the derivation of the estimates for the clock offset but, as a byproduct, the estimates of both fixed and variable link delays have also been obtained in (5.15) and (5.19), where their BLUE-OS again matches the MVUE. This outcome is also helpful since end-to-end delay measurements are frequently used in analyzing network performance and usually there is no provision inside the network to provide end-systems with information about the current status of the network. For example, packet delay statistics are important in examining the performance and reliability of the Internet, but it has no mechanism for providing feedback on network congestion to end-systems at the Internet Protocol (IP) layer. Moreover, these results are also useful for end-system protocols and applications that behave adaptively based on their control on the observed network performance. Lastly, the estimates of fixed and variable delays are also important in other areas such as continuous-media applications, e.g., audio and video applications need to absorb the delay jitter perceived at the receiver for smooth playout of the original stream (see [17] and [76]). Determining the correct amount of buffering and the reconstruction of the original timing play a vital role in obtaining better performance in such applications.

Chapter 6

CLOCK OFFSET AND SKEW ESTIMATION

We now turn our attention to a more accurate model defining the relationship between two clocks by the addition of clock skew. In practice, the time synchronization problem in WSNs generally involves two steps: synchronizing the nodes in the network to one common absolute time by adjusting clock phase offset (clock offset) among the nodes, and correcting the clock frequency offset (clock skew) relative to a certain standard frequency. The second step is required because the imperfections in quartz crystals and environmental conditions induce different clocks to run at slightly different frequencies. Actually, the effect of clock skew is the main reason why clock offset keeps drifting apart. Hence, adjusting clock skew guarantees long-term reliability of synchronization, and therefore reduces network-wide energy consumption in synchronization procedures. Indeed, developing long-term and network-wide time synchronization protocols that are energy-efficient represents one of the key strategies for the successful deployment of long-lived WSNs.

The main topics in this chapter are as follows. First, the MLE and the corresponding CRLB for the conventional clock offset model in a general sender–receiver protocol assuming a Gaussian model for the noise are derived. Second, the joint MLE and corresponding CRLB using a more realistic linear clock offset and skew model assuming Gaussian random delays are obtained. Third, the CRLB for the clock offset for the exponential delay model is derived as a performance threshold. Fourth, the joint MLE for the clock offset and skew under the exponential delay model is obtained and the corresponding algorithms to find these estimators are described in detail.

6.1 GAUSSIAN DELAY MODEL

As explained earlier, several pdf models have been proposed for random network delays. Even for an unknown delay distribution, the final error is a sum of many independent random components as described in Chapter 1. Exploiting the central limit theorem (CLT), which asserts that the pdf of the sum of a number of independent and iid random variables approaches that of a Gaussian random variable, the Gaussian model in our study is appropriate if the delays are thought to be the addition of a few such independent random processes. For example, suppose that the actual errors are uniformly distributed around a mean delay value, then the sum of just two such errors closely resembles the Gaussian pdf. In addition, the Gaussian distribution for the phase offset errors has been corroborated by a few authors, see, e.g., [22], based on laboratory measurements and tests. In this section, we derive the MLE and the CRLB for two scenarios. The first scenario deals with only the estimation of clock offset, while the second scenario addresses the problem of joint estimation of the clock offset and skew under the common assumption of Gaussian delays.

6.1.1 MAXIMUM LIKELIHOOD (ML) CLOCK OFFSET ESTIMATION

Assuming no clock skew at this stage and utilizing the same signaling mechanism as in Chapter 5 (see Figure 5.1), we compute the MLE and CRLB for the clock offset using the two-way timing message exchange model. Since the set of network delays $\{X_k\}_{k=1}^N$ and $\{Y_k\}_{k=1}^N$ are independently and normally distributed with the same mean μ and variance σ^2, the likelihood function based on the observations $\{U_k\}_{k=1}^N$ and $\{V_k\}_{k=1}^N$ is given by

$$\begin{aligned} L\left(\phi,\mu,\sigma^2\right) &= \left(2\pi\sigma^2\right)^{-N} \\ &\quad \times \exp\left\{-\frac{1}{2\sigma^2}\left[\sum_{k=1}^{N}\left(U_k-d-\phi-\mu\right)^2+\sum_{k=1}^{N}\left(V_k-d+\phi-\mu\right)^2\right]\right\}. \end{aligned}$$

Differentiating the log-likelihood function leads to

$$\begin{aligned} \frac{\partial \ln L\left(\phi\right)}{\partial\phi} &= -\frac{1}{2\sigma^2}\left[\sum_{k=1}^{N}\left(2\phi-2\left(U_k-d-\mu\right)\right)+\sum_{k=1}^{N}\left(2\phi+2\left(V_k-d-\mu\right)\right)\right] \\ &= -\frac{1}{\sigma^2}\left[\sum_{k=1}^{N}\left(2\phi-\left(U_k-V_k\right)\right)\right]. \end{aligned} \tag{6.1}$$

Hence, the MLE of clock offset is given by

$$\hat{\phi} = \arg\max_{\phi}\left[\ln L\left(\phi\right)\right] = \frac{\sum_{k=1}^{N}\left(U_k-V_k\right)}{2N} = \frac{\overline{U}-\overline{V}}{2}, \tag{6.2}$$

where $\overline{U}$ and $\overline{V}$ stand for the sample means of observations $\{U_k\}_{k=1}^N$ and $\{V_k\}_{k=1}^N$, respectively. Consequently, the MLE of clock offset can be obtained by dividing by a factor of 2 the difference between the means of observations $\{U_k\}_{k=1}^N$ and $\{V_k\}_{k=1}^N$.

6.1.2 Cramer–Rao Lower Bound (CRLB) for Clock Offset

The regularity conditions imposed for the existence of CRLB [44] hold for the given estimate since the expected value of (6.1) is 0. Thus, the CRLB for the MLE can be obtained by differentiating (6.1) with respect to ϕ, which leads to

$$\frac{\partial^2 \ln L(\phi)}{\partial \phi^2} = -\frac{2N}{\sigma^2}.$$

Hence, the CRLB for the clock offset is given by

$$\text{var}(\hat{\phi}) \geq -E\left[\frac{\partial^2 \ln L(\phi)}{\partial \phi^2}\right]^{-1} = \frac{\sigma^2}{2N}. \tag{6.3}$$

6.1.3 Joint Maximum Likelihood Estimation (JMLE) of Clock Offset and Skew

Since every oscillator has its unique clock frequency, the clock offset between two nodes generally keeps increasing. Therefore, a fixed value model for clock time difference as assumed in the previous section might not be sufficient for certain practical situations. Hence, estimating the difference in clock frequencies between two nodes (i.e., clock skew) increases synchronization accuracy and guarantees long-term reliability. In this section, we derive the JMLE for clock offset and skew based on the two-way timing message exchange model with Gaussian delays.

The theory applied thus far for finding the MLE and CRLB for the clock offset (assuming no clock skew) can be extended to find the JMLE and CRLB for a more general clock model. Figure 6.1 shows the effect of clock offset (ϕ) and skew (ω) on timing message exchanges between two nodes. Here, timestamps in the kth message exchange $T_{1,k}$ and $T_{4,k}$ are measured by the local clock of node A, and $T_{2,k}$ and $T_{3,k}$ are measured by the local clock of node B, respectively. Node A transmits a synchronization packet, containing the level and identity of node A and the value of timestamp $T_{1,k}$, to node B. Node B receives it at $T_{2,k}$ and transmits an acknowledgement packet to node A at $T_{3,k}$. This packet contains the level and identity of node B and the value of timestamps $T_{1,k}$, $T_{2,k}$, and $T_{3,k}$. Then node A finally receives the packet at $T_{4,k}$.

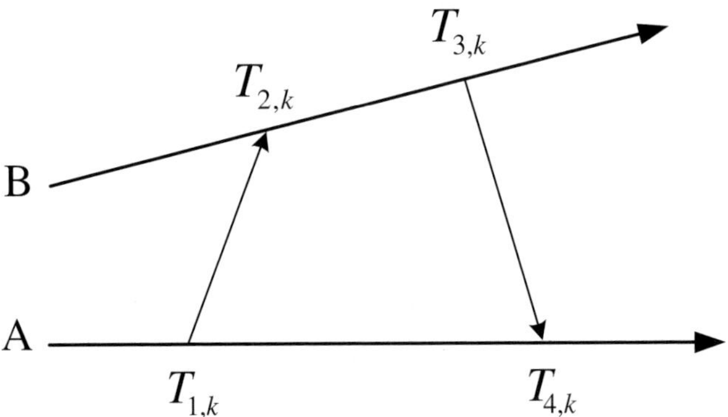

Figure 6.1: Two-way timing message exchange model in the presence of clock offset and skew.

Taking $T_{1,1}$ as the reference time ($T_{1,1} = 0$), the timestamp at node B in the kth uplink message $T_{2,k}$, is given by

$$T_{2,k} = (T_{1,k} + d + X_k)\,\omega + \phi, \tag{6.4}$$

and the timestamp at node B in the kth downlink message $T_{3,k}$, is represented by

$$T_{3,k} = (T_{4,k} - d - Y_k)\omega + \phi. \tag{6.5}$$

Assuming $\{X_k\}_{k=1}^N$ and $\{Y_k\}_{k=1}^N$ are zero mean independent Gaussian distributed RVs with variance σ^2, then the joint pdf of $\mathbf{X} \triangleq \{X_k\}_{k=1}^N$ and $\mathbf{Y} \triangleq \{Y_k\}_{k=1}^N$ is given by

$$\begin{aligned} &f_{\mathbf{X},\mathbf{Y}}(\mathbf{x},\mathbf{y}) \\ &= \left(2\pi\sigma^2\right)^{-N} \exp\left\{-\frac{1}{2\sigma^2}\sum_{k=1}^N \left[\left(\frac{T_{2,k}-\phi}{\omega} - T_{1,k} - d\right)^2 + \left(T_{4,k} - d - \frac{T_{3,k}-\phi}{\omega}\right)^2\right]\right\}. \end{aligned}$$

Further assuming that the fixed portion of delay d is known and introducing the new variable $\omega' \triangleq 1/\omega$, the likelihood function for $(\phi, \omega', \sigma^2)$, based on the observations $\{T_{1,k}\}_{k=1}^N$, $\{T_{2,k}\}_{k=1}^N$, $\{T_{3,k}\}_{k=1}^N$, and $\{T_{4,k}\}_{k=1}^N$, is given by

$$\begin{aligned} &L\left(\phi,\omega',\sigma^2\right) = \left(2\pi\sigma^2\right)^{-N} \\ &\times \exp\left(-\frac{1}{2\sigma^2}\sum_{k=1}^N \left\{\left[\omega'(T_{2,k}-\phi) - (T_{1,k}+d)\right]^2 + \left[\omega'(\phi - T_{3,k}) + (T_{4,k}-d)\right]^2\right\}\right). \end{aligned}$$

Differentiating the log-likelihood function with respect to ϕ leads to

$$\frac{\partial \ln L\left(\phi,\omega',\sigma^2\right)}{\partial\phi} = -\frac{1}{\sigma^2}\sum_{k=1}^N \left[\omega'^2\left(2\phi - T_{2,k} - T_{3,k}\right) + \omega'\left(T_{1,k} + T_{4,k}\right)\right]. \tag{6.6}$$

Hence, in the given clock skew model, the MLE of clock offset $\hat{\phi}$ is expressed as

$$\hat{\phi} = \frac{\sum_{k=1}^{N} \left[\hat{\omega}' \left(T_{2,k} + T_{3,k}\right) - \left(T_{1,k} + T_{4,k}\right)\right]}{2N\hat{\omega}'}. \tag{6.7}$$

Note that, for the clock skew model with Gaussian random delays, there is an additional term which depends on $\hat{\omega}$, and this result reduces to (6.2) when $\hat{\omega}$ is 1. Similarly, differentiating the log-likelihood function with respect to ω' leads to

$$\begin{aligned}\frac{\partial \ln L\left(\phi, \omega', \sigma^2\right)}{\partial \omega'} &= -\frac{1}{\sigma^2}\left\{\sum_{k=1}^{N} \omega' \left[(T_{2,k} - \phi)^2 + (T_{3,k} - \phi)^2\right]\right. \\ &\quad \left. - \sum_{k=1}^{N} \left[(T_{1,k} + d)(T_{2,k} - \phi) + (T_{4,k} - d)(T_{3,k} - \phi)\right]\right\}. \end{aligned} \tag{6.8}$$

Thus, the estimate $\hat{\omega}'$ maximizing the log-likelihood function is given by

$$\hat{\omega}' = \frac{\sum_{k=1}^{N} \left[(T_{1,k} + d)(T_{2,k} - \hat{\phi}) + (T_{4,k} - d)(T_{3,k} - \hat{\phi})\right]}{\sum_{k=1}^{N} \left[(T_{2,k} - \hat{\phi})^2 + (T_{3,k} - \hat{\phi})^2\right]}.$$

Hence, the JMLE of clock skew $\hat{\omega}$ is given by

$$\hat{\omega} = \frac{\sum_{k=1}^{N} \left[(T_{2,k} - \hat{\phi})^2 + (T_{3,k} - \hat{\phi})^2\right]}{\sum_{k=1}^{N} \left[(T_{1,k} + d)(T_{2,k} - \hat{\phi}) + (T_{4,k} - d)(T_{3,k} - \hat{\phi})\right]}. \tag{6.9}$$

In the sequel, the JMLE of ϕ and ω can be obtained by plugging the expression of $\hat{\phi}$ (6.7) into that of $\hat{\omega}$ (6.9), which gives

$$\hat{\omega} = \frac{\sum\limits_{k=1}^{N} (T_{2,k} + T_{3,k}) - 2N\hat{\phi}}{\sum\limits_{k=1}^{N} (T_{1,k} + T_{4,k})} = \frac{\sum\limits_{k=1}^{N} \left[(T_{2,k} - \hat{\phi})^2 + (T_{3,k} - \hat{\phi})^2\right]}{\sum\limits_{k=1}^{N} \left[(T_{1,k} + d)(T_{2,k} - \hat{\phi}) + (T_{4,k} - d)(T_{3,k} - \hat{\phi})\right]}. \tag{6.10}$$

After some manipulations, the JMLE of clock offset and skew is given by

$$\hat{\phi}_{GML} = \frac{\sum\limits_{k=1}^{N} (T_{1,k} + T_{4,k}) \sum\limits_{k=1}^{N} (T_{2,k}^2 + T_{3,k}^2) - \sum\limits_{k=1}^{N} (T_{2,k} + T_{3,k})Q}{\sum\limits_{k=1}^{N} (T_{2,k} + T_{3,k}) \sum\limits_{k=1}^{N} (T_{1,k} + T_{4,k}) - 2NQ}, \tag{6.11}$$

$$\hat{\omega}_{GML} = \frac{-2N\left[\sum_{k=1}^{N}(T_{1,k}+T_{4,k})\sum_{k=1}^{N}(T_{2,k}^2+T_{3,k}^2) - Q\sum_{k=1}^{N}(T_{2,k}+T_{3,k})\right]}{\sum_{k=1}^{N}(T_{1,k}+T_{4,k})\left[\sum_{k=1}^{N}(T_{2,k}+T_{3,k})\sum_{k=1}^{N}(T_{1,k}+T_{4,k}) - 2NQ\right]} + \frac{\sum_{k=1}^{N}(T_{2,k}+T_{3,k})}{\sum_{k=1}^{N}(T_{1,k}+T_{4,k})}, \tag{6.12}$$

where $Q \triangleq \sum_{k=1}^{N}\left[T_{1,k}T_{2,k} + T_{3,k}T_{4,k} + (T_{2,k} - T_{3,k})d\right]$. Note that the JMLE depends on the value of the fixed portion of delays d by means of the factor Q, which is assumed to be known in this section. Although in certain situations, d is known (e.g., $d = 0$ for small-scale WSNs with RF transmissions), we do not consider d as another unknown (nuisance) parameter due to the fact that if it is the resulting estimation problem becomes highly non-linear and complex. In Chapter 7, we will consider the design of simplified but suboptimal estimators that overcome the complex problem of estimating d.

Finally, notice that some authors (see, e.g., [22], [58]) use a linear regression model to capture the relationship between the clocks of two nodes, i.e., (eliminating known d)

$$T_{2,k} = T_{1,k}\omega + \phi + X_k, \quad k = 1, \ldots, N, \tag{6.13}$$

which is a restrictive assumption, since ω has its effect on the clock of the second node during the message delay until the message is timestamped on its reception. This is particularly important in networks where the message delays are large, e.g., underwater acoustic networks. We assert that using the more realistic model described by (6.4) and (6.5) results in superior performance of the synchronization protocol. To show the effect on MSE of using the more realistic model, we have simulated the MSE of two clock offset estimators, one derived based on the correct model and the other derived using the standard least squares solution based on linear regression (6.13); these are shown in Figure 6.2. It is clear that using the more realistic model results in a lower MSE of the clock offset estimate.

6.1.4 Cramer–Rao Lower Bound (CRLB) for Clock Offset and Skew

The CRLB for the vector parameter $\boldsymbol{\theta} = [\phi, \omega]^T$ can be derived from the 2×2 Fisher information matrix $\mathbf{I}(\boldsymbol{\theta})$ by taking its inverse [44]. From (6.6) and (6.8), the second-order derivatives of the log-likelihood function with respect to ϕ and ω' are

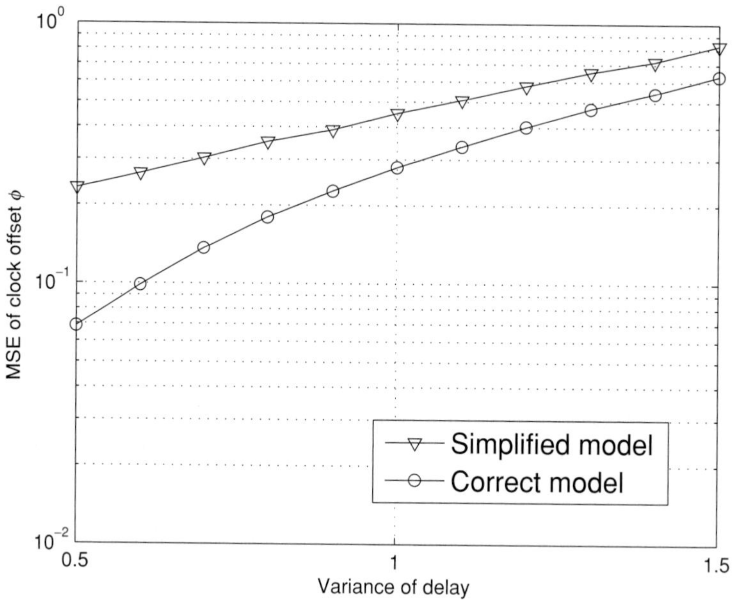

Figure 6.2: MSE of clock offset estimate $\hat{\phi}$ as a function of delay variance.

found as

$$\begin{aligned}
\frac{\partial^2 \ln L\left(\phi, \omega', \sigma^2\right)}{\partial \phi^2} &= -\frac{2N\omega'^2}{\sigma^2}, \\
\frac{\partial^2 \ln L\left(\phi, \omega', \sigma^2\right)}{\partial \omega'^2} &= -\frac{1}{\sigma^2}\sum_{k=1}^{N}\left[(T_{2,k}-\phi)^2 + (T_{3,k}-\phi)^2\right], \\
\frac{\partial^2 \ln L\left(\phi, \omega', \sigma^2\right)}{\partial \phi \omega'} &= -\frac{1}{\sigma^2}\sum_{k=1}^{N}\left(2\omega'\phi - \omega' T_{2,k} + T_{1,k} - \omega T_{3,k} - T_{4,k}\right).
\end{aligned}$$

Taking the negative expectations yields

$$\begin{aligned}
-E\left[\frac{\partial^2 \ln L\left(\phi, \omega', \sigma^2\right)}{\partial \phi^2}\right] &= \frac{2N\omega'^2}{\sigma^2}, \\
-E\left[\frac{\partial^2 \ln L\left(\phi, \omega', \sigma^2\right)}{\partial \omega'^2}\right] &= \frac{1}{\sigma^2}\sum_{k=1}^{N} E_{X_k,Y_k}\left[\frac{(X_k + T_{1,k} + d)^2 + (Y_k - T_{4,k} + d)^2}{\omega'^2}\right] \\
&\stackrel{(a)}{=} \frac{\sum_{k=1}^{N}\left((T_{1,k}+d)^2 + (T_{4,k}-d)^2 + 2\sigma^2\right)}{\sigma^2\omega'^2}, \\
-E\left[\frac{\partial^2 \ln L\left(\phi, \omega', \sigma^2\right)}{\partial \phi \omega'}\right] &= -\frac{1}{\sigma^2}\sum_{k=1}^{N} E_{X_k,Y_k}\left[2\omega'\left(2\phi - T_{2,k} - T_{3,k}\right) + T_{1,k} + T_{4,k}\right] \\
&\stackrel{(b)}{=} \frac{N}{\sigma^2}\left(\overline{T_1} + \overline{T_4}\right),
\end{aligned}$$

where (a) and (b) are due to $X_k = \omega'(T_{2,k} - \phi) - (T_{1,k} + d)$ and $Y_k = \omega'(\phi - T_{3,k}) + (T_{4,k} - d)$. Therefore, the Fisher information matrix becomes

$$\begin{aligned} \mathbf{I}(\boldsymbol{\theta}) &= \begin{bmatrix} -E\left[\frac{\partial^2 \ln L(\phi,\omega',\sigma^2)}{\partial\phi^2}\right] & -E\left[\frac{\partial^2 \ln L(\phi,\omega',\sigma^2)}{\partial\phi\omega'}\right] \\ -E\left[\frac{\partial^2 \ln L(\phi,\omega',\sigma^2)}{\partial\omega'\phi}\right] & -E\left[\frac{\partial^2 \ln L(\phi,\omega',\sigma^2)}{\partial\omega'^2}\right] \end{bmatrix} \\ &= \frac{1}{\sigma^2}\begin{bmatrix} 2N\omega^2 & N\left(\overline{T_1}+\overline{T_4}\right) \\ N\left(\overline{T_1}+\overline{T_4}\right) & \frac{1}{\omega'^2}\sum_{k=1}^{N}\left[(T_{1,k}+d)^2+(T_{4,k}-d)^2+2\sigma^2\right] \end{bmatrix}. \end{aligned} \tag{6.14}$$

Now the CRLB can be obtained by taking the inverse of the $[k,k]$th element of the Fisher information matrix (i.e., $\text{var}(\hat{\theta}_k) \geq \left[\mathbf{I}^{-1}(\boldsymbol{\theta})\right]_{ii}$), and the inverse $\mathbf{I}^{-1}(\boldsymbol{\theta})$ is given by

$$\mathbf{I}^{-1}(\boldsymbol{\theta}) = \sigma^2 \begin{bmatrix} \frac{V}{\omega'^2 N\left[2V - N\left(\overline{T_1}+\overline{T_4}\right)^2\right]} & \frac{-\left(\overline{T_1}+\overline{T_4}\right)}{2V-N\left(\overline{T_1}+\overline{T_4}\right)^2} \\ \frac{-\left(\overline{T_1}+\overline{T_4}\right)}{2V-N\left(\overline{T_1}+\overline{T_4}\right)^2} & \frac{2\omega'^2}{2V-N\left(\overline{T_1}+\overline{T_4}\right)^2} \end{bmatrix}, \tag{6.15}$$

where $V = \sum_{k=1}^{N}\left[(T_{1,k}+d)^2+(T_{4,k}-d)^2+2\sigma^2\right]$. Consequently, from the result in [44], the CRLBs of clock offset and skew for the Gaussian delay model are respectively given by

$$\text{var}(\hat{\phi}^{GML}) \geq \frac{\sigma^2\omega^2 V}{N\left[2V - N\left(\overline{T_1}+\overline{T_4}\right)^2\right]}, \tag{6.16}$$

$$\begin{aligned} \text{var}(\hat{\omega}^{GML}) &\geq \left(\frac{\partial\omega}{\partial\omega'}\right)^2 \frac{2\sigma^2\omega'^2}{2V - N\left(\overline{T_1}+\overline{T_4}\right)^2} \\ &= \frac{2\sigma^2\omega^2}{2V - N\left(\overline{T_1}+\overline{T_4}\right)^2}. \end{aligned} \tag{6.17}$$

6.2 Exponential Delay Model

A detailed justification of modeling the network delays as coming from an exponential distribution was presented in Chapter 5. Since the MLE for the clock offset under exponential delays has been derived in [41] and already mentioned in (5.1), we derive the corresponding CRLB for the clock offset in the next section. Afterwards, the joint MLE for both the clock offset and skew is obtained and the corresponding algorithms for finding those estimates are also presented.

6.2.1 CRAMER–RAO LOWER BOUND (CRLB) FOR CLOCK OFFSET

It was proven in [41] that the MLE of ϕ exists when d is unknown and exhibits the same form as the estimator proposed in [71], which is given by

$$\hat{\phi} = \frac{U_{(1)} - V_{(1)}}{2}, \tag{6.18}$$

where N stands for the number of observations of delay measurements and the subscript (1) denotes the first-order statistic of the corresponding data set. In this section, we proceed towards obtaining the CRLB for this clock offset under the exponential delay model.

Note that (6.18) can be rewritten as

$$\hat{\phi} = \frac{U_{(1)} - V_{(1)}}{2} = \phi + \frac{X_{(1)} - Y_{(1)}}{2},$$

where $X_{(1)}$ and $Y_{(1)}$ denote the corresponding order statistics of $\{X_k\}_{k=1}^N$ and $\{Y_k\}_{k=1}^N$, respectively. Define the new variable $Z \triangleq X_{(1)} - Y_{(1)}$, then the pdf of Z can be found as follows. Since the order statistics $X_{(1)}$ and $Y_{(1)}$ are independent, the pdf of RV Z, $f_Z(z)$, can be found by transforming the joint distribution of RVs $X_{(1)}$ and $Y_{(1)}$ using the dummy variable $S = Y_{(1)}$. From the modeling assumptions, the pdfs of the uplink and downlink delays, X_k and Y_k, are given respectively by

$$f_{X_k}(x) = \frac{1}{\lambda_1} \exp\left(-\frac{x}{\lambda_1}\right) \qquad x \geq 0,$$
$$f_{Y_k}(y) = \frac{1}{\lambda_2} \exp\left(-\frac{y}{\lambda_2}\right) \qquad y \geq 0.$$

It is well known that the pdfs of the order statistics $X_{(1)}$ and $Y_{(1)}$ are given by

$$f_{X_{(1)}}(x) = N\left(1 - F_{X_i}(x)\right)^{N-1} f_{X_i}(x) = \frac{N}{\lambda_1} \exp\left(-\frac{N}{\lambda_1} x\right) \qquad x \geq 0,$$
$$f_{Y_{(1)}}(x) = N\left(1 - F_{Y_i}(y)\right)^{N-1} f_{Y_i}(y) = \frac{N}{\lambda_2} \exp\left(-\frac{N}{\lambda_2} y\right) \qquad y \geq 0.$$

Since the *Jacobian* of this transformation is 1, a joint distribution of RVs Z and S is given by

$$\begin{aligned} f_{Z,S}(z,s) &= f_{X_{(1)},Y_{(1)}}(z+s,s) = f_{X_{(1)}}(z+s)\, f_{Y_{(1)}}(s) \\ &= \frac{N^2}{\lambda_1 \lambda_2} \exp\left(-\frac{N}{\lambda_1} z\right) \exp\left(-N\left(\frac{\lambda_1+\lambda_2}{\lambda_1\lambda_2}\right) s\right) \quad z \geq -s,\ s \geq 0. \end{aligned} \tag{6.19}$$

Integrating (6.19) with respect to s yields

$$f_Z(z) = \begin{cases} \frac{N}{(\lambda_1+\lambda_2)} \exp\left(-\frac{N}{\lambda_1} z\right) & z > 0 \\ \frac{N}{(\lambda_1+\lambda_2)} \exp\left(\frac{N}{\lambda_2} z\right) & z < 0 \end{cases}. \tag{6.20}$$

Let $W \triangleq U_{(1)} - V_{(1)}$, then the pdf of W as a function of ϕ is given by

$$f_w(w;\phi) = \begin{cases} \frac{N}{(\lambda_1+\lambda_2)} \exp\left(-\frac{N}{\lambda_1}(w-2\phi)\right) & w > 2\phi \\ \frac{N}{(\lambda_1+\lambda_2)} \exp\left(\frac{N}{\lambda_2}(w-2\phi)\right) & w < 2\phi \end{cases}. \tag{6.21}$$

Note that the estimate $\hat{\phi}$ is biased when uplink and downlink delays are asymmetrically distributed, i.e., when $\lambda_1 \neq \lambda_2$. Thus, to derive the CRLB for the estimator, the delays are assumed to be symmetric, which yields $\lambda_1 = \lambda_2 = \alpha$. Now (6.21) can be rewritten as

$$f_W(w;\phi) = \frac{N}{2\alpha} e^{-\frac{N}{\alpha}|w-2\phi|}.$$

Differentiating the logarithm of (6.21) with respect to ϕ gives

$$\frac{\partial \ln f_W(w;\phi)}{\partial \phi} = \begin{cases} \frac{2N}{\alpha} & w > 2\phi \\ -\frac{2N}{\alpha} & w < 2\phi \end{cases}, \tag{6.22}$$

where the regularity condition of the CRLB holds since (6.22) is finite and the expected value of (6.22) is 0. Calculating the expected value of the square of (6.22) yields

$$E\left[\left(\frac{\partial \ln f_W(w;\phi)}{\partial \phi}\right)^2\right] = \frac{4N^2}{\alpha^2}.$$

Therefore, the CRLB of clock offset, $\hat{\phi}$, is given by

$$\operatorname{var}(\hat{\phi}) \geq E\left[\left(\frac{\partial \ln f_W(w;\phi)}{\partial \phi}\right)^2\right]^{-1} = \frac{\alpha^2}{4N^2}. \tag{6.23}$$

Figure 6.3 shows the simulation results corresponding to the variance and CRLB of the MLE when α is 1. It can be seen that the variance of the estimate goes to zero as N increases (asymptotically efficient), and is proportional to α^2.

In Figure 6.4, the variances of both MLEs are compared in exponential and normal random delay channels. It can be seen that the performance of the ML clock offset estimator is strongly dependent on the type of random delay model. Figure 6.4

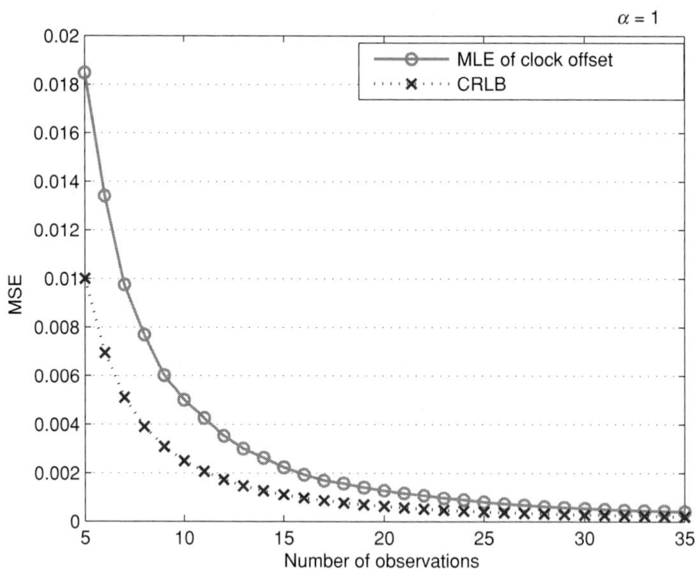

Figure 6.3: CRLB and variance of the MLE of clock offset for the exponential delay model ($\alpha = 1$).

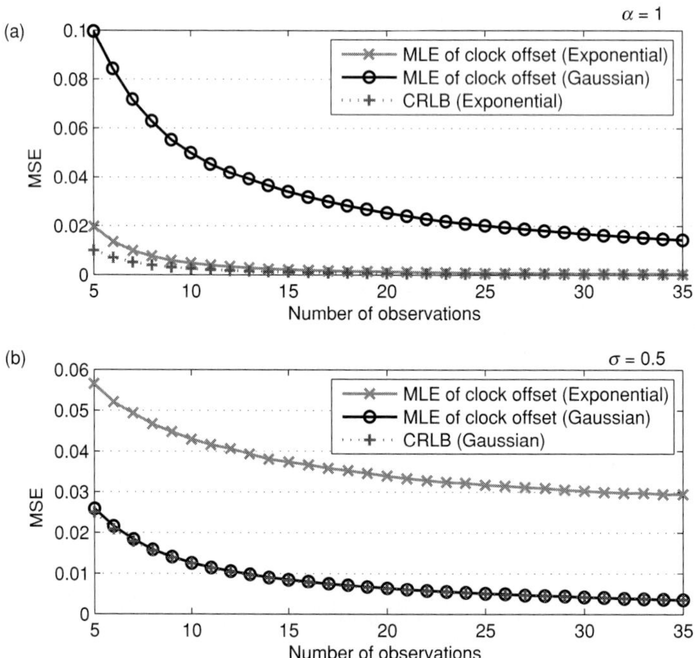

Figure 6.4: Variances of MLEs of clock offset for (a) exponential and (b) Gaussian delays ($\alpha = 1$ and $\sigma = 0.5$).

illustrates that an MLE derived assuming Gaussian noise does not perform well in the presence of network delays that are exponentially distributed. At the same time, an MLE derived assuming the exponential pdf model for the noise performs poorly in the presence of normally distributed noise. These considerations motivate the need to build clock offset estimators that are robust with respect to the distribution of network delays, which in general cannot be predicted exactly and depend on a number of factors. The problem of designing robust clock offset estimators will be addressed in Chapter 14.

6.2.2 JOINT MAXIMUM LIKELIHOOD ESTIMATION (JMLE) OF CLOCK OFFSET AND SKEW

From the timing message exchange model described in (6.4) and (6.5) but with exponential delays, the general form of the likelihood function is given by

$$\begin{aligned} L(\alpha, d, \omega, \phi) &= \alpha^{-2N} \exp\left[-\frac{1}{\alpha}\left\{\sum_{k=1}^{N} \frac{T_{2,k} - T_{3,k}}{\omega} - \sum_{k=1}^{N} (T_{1,k} - T_{4,k}) - 2Nd\right\}\right] \\ &\times \prod_{k=1}^{N} I\left[\frac{T_{2,k} - \phi}{\omega} - T_{1,k} - d \geq 0; T_{4,k} - \frac{T_{3,k} - \phi}{\omega} - d \geq 0\right], \end{aligned} \tag{6.24}$$

where the indicator function $I[\cdot]$ is defined as

$$I[x \geq 0] = \begin{cases} 1, & x \geq 0 \\ 0, & x < 0 \end{cases}.$$

Note that d is always positive since it represents the fixed delay, while ω is also always positive because it has been realistically assumed that none of the clocks is either standing still ($\omega = 0$) or running backward ($\omega < 0$). An ideal value of $\omega = 1$ means that the clock is running at the standard rate. Also, notice that when $\omega = 1$, the MLE of clock offset ϕ was derived in [41] and takes the form

$$\hat{\phi} = \frac{1}{2}\left[\min_{1 \leq k \leq N}(T_{2,k} - T_{1,k}) - \min_{1 \leq k \leq N}(T_{4,k} - T_{3,k})\right]. \tag{6.25}$$

From here onwards, without any loss of generalization, we will assume that α is known. This is because even if α is unknown, due to the form of the reduced likelihood function $L(d, \omega, \phi)$ as shown in [41], the MLE $(\hat{d}, \hat{\omega}, \hat{\phi})$ remains the same. When $\omega \neq 1$, in maximizing the likelihood for this model over the set $\Theta = \{(d, \phi, \omega) : d > 0, -\infty < \phi < \infty, \omega > 0\}$, four different cases will be considered:

Case I: fixed delay d known, clock offset ϕ known;

Case II: fixed delay d unknown, clock offset ϕ known;

Case III: fixed delay d known, clock offset ϕ unknown;

Case IV: fixed delay d unknown, clock offset ϕ unknown.

An important remark needs to be made here. A preliminary examination of Cases I and II (i.e., when ϕ is known) is necessary because it gives insight into the shape of the support region over which the likelihood function is non-zero. As is the case with exponential models, the MLEs for the location parameters are found by taking effectively into account the boundary conditions. For the first two cases, the support of the likelihood region is a two-dimensional region and it is relatively easy to find the parameters on the boundary maximizing the likelihood function. Finding the MLEs for Cases III and IV (i.e., when ϕ is unknown) requires the visualization of the likelihood function support region in three dimensions, and obtaining a somewhat primitive knowledge of the two-dimensional support region for the likelihood function in Cases I and II greatly helps in preparing our intuition and solving the more complex three-dimensional optimization problem. Therefore, we next proceed with a stepwise approach by considering these four cases separately one-by-one.

Case I: fixed delay d known, clock offset ϕ known

Without any loss of generalization, the likelihood function in this case can be obtained by making $\phi = 0$ in (6.24). From the form of the likelihood function, we can see that it is non-zero only over a certain support region defined by the limits of the indicator function $I[\cdot]$. Since d is fixed and known, the set of constraints in (6.24), namely $d > 0$, $\omega > 0$ and

$$d \leq \frac{T_{2,k}}{\omega} - T_{1,k}, \quad k = 1, \ldots, N, \tag{6.26}$$

$$d \leq T_{4,k} - \frac{T_{3,k}}{\omega}, \quad k = 1, \ldots, N, \tag{6.27}$$

can be equivalently put in the form

$$\begin{aligned} d &> 0, \ \omega > 0, \\ \frac{T_{3,k}}{T_{4,k} - d} \leq \ &\omega \ \leq \frac{T_{2,k}}{T_{1,k} + d}, \quad k = 1, \ldots, N. \end{aligned} \tag{6.28}$$

Figure 6.5 shows various upper bounds (6.26) and (6.27) of the likelihood support region in the plane (d, ω), and the bold line is the region over which the likelihood function has to be maximized. It is evident from the figure that for a known fixed d, the likelihood function depends on the unknown ω only and is maximized by

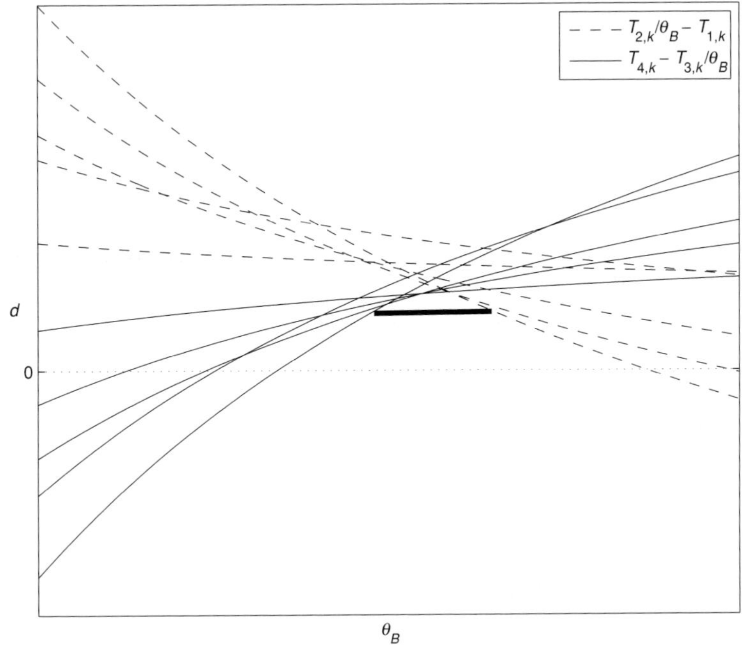

Figure 6.5: The support region of the likelihood function is shown as a bold line in the (d, ω) plane together with the upper bounds (6.26) and (6.27).

taking ω as small as possible. This is because the factor $\sum_{k=1}^{N}(T_{2,k} - T_{3,k})$ in (6.24) is always negative. Therefore, the smallest value of ω over the bold line, as shown in Figure 6.5, is the MLE $\hat{\omega}$, which coincides with one of the curves $d = T_{4,k} - T_{3,k}/\omega$, $k = 1, \ldots, N$. Let j $(1 \leq j \leq N)$ denote the index of the curve on which the MLE is achieved. Thus, from (6.28), $j = \arg\max_k \{T_{3,k}/(T_{4,k} - d)\}$ and

$$\hat{\omega} = \frac{T_{3,j}}{T_{4,j} - d}.$$

The index j, which gives the set of timestamps $\{T_{3,j}, T_{4,j}\}$ required for finding the MLE, is the one which gives the minimum possible $\hat{\omega}$ over the allowable region. Since d is known, we can find j, and hence the corresponding $\hat{\omega}$, by Algorithm 1.

Algorithm 1: Finding $\hat{\omega}$ for d known, ϕ known

---

1. Find $\omega_k = \frac{T_{3,k}}{T_{4,k} - d}$, for $k = 1, \ldots, N$;
2. $j = \arg\max_k \{\omega_k\}$;
3. $\hat{\omega} = \hat{\omega}_j$;

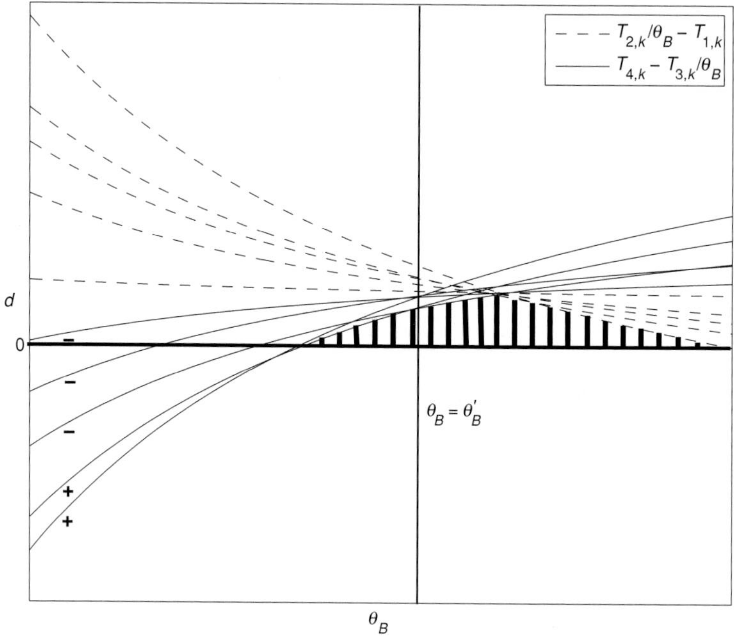

Figure 6.6: The support region of the likelihood function is shown as the hatched area in the (d, ω) plane, together with the upper bounds (6.29) and (6.30). Also shown to the left in this figure is the sign of the term $2NT_{3,j} - \sum_{k=1}^{N}(T_{3,k} - T_{2,k})$ for each $j = 1, \ldots, N$.

Algorithm 1 utilizes the fact that the solid line cuts all the curves $T_{4,k} - T_{3,k}/\omega$, $k = 1, \ldots, N$ but the likelihood function is zero beyond its intersection with the first curve, which is the maximum of these intersections and therefore gives the MLE. For this approach, a total of N values needs to be compared. To simplify the exposition, in what follows we will use the terminology *the curves* $T_{4,k} - T_{3,k}/\omega$, $k = 1, \ldots, N$, instead of *the curves* $d = T_{4,k} - T_{3,k}/\omega$, $k = 1, \ldots, N$.

Case II: fixed delay d unknown, clock offset ϕ known

The likelihood function in this case is similar to that in Case I, but with one major difference: the fixed delay d is unknown. The hatched region in Figure 6.6 is the subset of Θ over which the likelihood function is non-zero. It can be described in terms of the following constraints:

$$d > 0, \quad \omega > 0,$$

$$d \leq \frac{T_{2,k}}{\omega} - T_{1,k}, \quad k = 1, \ldots, N, \tag{6.29}$$

$$d \leq T_{4,k} - \frac{T_{3,k}}{\omega}, \quad k = 1, \ldots, N. \tag{6.30}$$

This likelihood function in (6.24) is maximized by making its argument,

$$\xi = \sum_{k=1}^{N} \frac{T_{2,k} - T_{3,k}}{\omega} - \sum_{k=1}^{N} (T_{1,k} - T_{4,k}) - 2Nd, \tag{6.31}$$

as small as possible. Although Figure 6.6 shows only the support region and not the likelihood function itself, ξ can be linked to this figure by rewriting it in the form

$$\xi = \sum_{k=1}^{N} \left(\frac{T_{2,k}}{\omega} - T_{1,k} \right) + \sum_{k=1}^{N} \left(T_{4,k} - \frac{T_{3,k}}{\omega} \right) - 2Nd,$$

and noting that for any $\omega = \omega'$, ξ is the sum of the ordinates of all points on the curves $(T_{2,k}/\omega - T_{1,k})$, $k = 1, \ldots, N$, and $(T_{4,k} - T_{3,k}/\omega)$, $k = 1, \ldots, N$, intercepting the vertical line $\omega = \omega'$, minus $2N$ times $\hat{d}$ (which is the intersection of $\omega = \omega'$ with either $\min_{1 \le k \le N} (T_{2,k}/\omega - T_{1,k})$ or $\min_{1 \le k \le N} (T_{4,k} - T_{3,k}/\omega)$ as proved in Lemma 6.1 below). Utilizing the fact that ξ depends on two parameters, ω and d, we will now derive the MLE with the help of the following four lemmas:

Lemma 6.1 *The MLE $\hat{d}$ lies on either $\min_{1 \le k \le N} (T_{2,k}/\omega - T_{1,k})$ or $\min_{1 \le k \le N} (T_{4,k} - T_{3,k}/\omega)$, i.e., on the boundary of the support region.*

Proof. This can be proved by contradiction. Let us assume that the $\hat{d}$ does not lie on the boundary, but somewhere else inside the support region. Then for some minimizing $\hat{\omega}$, ξ can be further decreased by increasing $\hat{d}$ to the top of the allowable region (which coincides with one of the above mentioned curves) for the same $\hat{\omega}$, hence a contradiction.

Lemma 6.2 *The MLE $\hat{d}$ lies either on the uppermost vertex formed by the intersection of the curves $\min_{1 \le k \le N} (T_{2,k}/\omega - T_{1,k})$ and $\min_{1 \le k \le N} (T_{4,k} - T_{3,k}/\omega)$ (shown as point A in Figure 6.7) or on one of the vertices formed by the intersection of the curves $(T_{4,k} - T_{3,k}/\omega)$, $k = 1, \ldots, N$ (shown as points B, C, etc., in Figure 6.7).*

Proof. It is straightforward to see from (6.31) that when $T_{2,k} = T_{3,k}$, for all k, ξ can be minimized by making d as large as possible, which is the intersection of the curves $\min_{1 \le k \le N} (T_{2,k}/\omega - T_{1,k})$ and $\min_{1 \le k \le N} (T_{4,k} - T_{3,k}/\omega)$. Hence, the MLE $(\hat{d}, \hat{\omega})$ is

$$\hat{d} = \frac{T_{2,i} T_{4,j} - T_{1,i} T_{3,j}}{T_{2,i} + T_{3,j}}, \tag{6.32}$$

and

$$\hat{\omega} = \frac{T_{2,i} + T_{3,j}}{T_{1,i} + T_{4,j}}, \tag{6.33}$$

where the i, j represent the indices of $\min_{1 \le k \le N} (T_{2,k}/\omega - T_{1,k})$ and $\min_{1 \le k \le N}(T_{4,k} - T_{3,k}/\omega)$, respectively, intersecting at the maximum $\hat{d}$ (which is the uppermost vertex

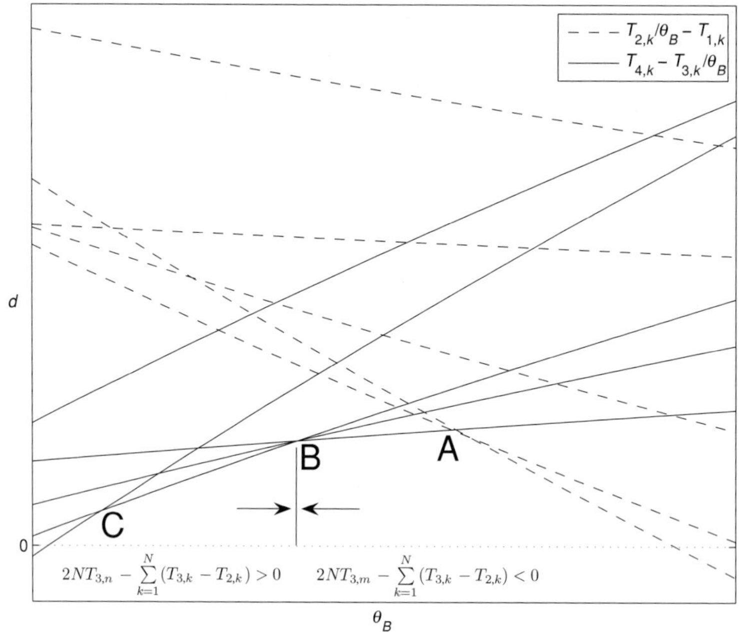

Figure 6.7: A close-up of the support region of the likelihood function in the (d, ω) plane together with upper bounds (6.29) and (6.30).

shown as point A in Figure 6.7). Note that in order to find this MLE, a total of N^2 intersections have to be compared.

When $T_{2,k} \neq T_{3,k}$, for some k, the problem becomes a little involved. From Lemma 6.1, we know that $\hat{d}$ lies somewhere on the boundary of the support region. Notice further that according to (6.31) in order to minimize ξ it is necessary to select d as large as possible and ω as small as possible.

Suppose that $\hat{d}$ lies on $\min_{1\leq k\leq N}(T_{2,k}/\omega - T_{1,k})$ and let $i = \arg\min_{1\leq k\leq N}(T_{2,k}/\omega - T_{1,k})$ corresponding to the maximum d (i.e., point A in Figure 6.7), then from (6.31) ξ can be expressed as

$$
\begin{aligned}
\xi &= \sum_{k=1}^{N} \frac{T_{2,k} - T_{3,k}}{\omega} - \sum_{k=1}^{N}(T_{1,k} - T_{4,k}) - 2N\left(\frac{T_{2,i}}{\omega} - T_{1,i}\right) \\
&= \frac{1}{\omega}\sum_{k=1}^{N}(T_{2,k} - T_{3,k} - 2T_{2,i}) - \sum_{k=1}^{N}(T_{1,k} - T_{4,k} - 2T_{1,i}). \qquad (6.34)
\end{aligned}
$$

Since the term $\sum_{k=1}^{N}(T_{2,k} - T_{3,k} - 2T_{2,i})$ is always negative, ξ can be minimized by taking ω as small as possible on $\min_{1\leq k\leq N}(T_{2,k}/\omega - T_{1,k})$. Hence, $\hat{d}$ and $\hat{\omega}$ in this general case are *equal to* or *less than* the MLE given by (6.32) and (6.33), respectively (i.e., either on point A shown in Figure 6.7 or to the left of it). An alternative justification for the fact that $\hat{d}$ and $\hat{\omega}$ are *equal to* or *less than* the MLE expressions given by (6.32) and (6.33), respectively, is to assume by contradiction that $\hat{d}$ lies on $\min_{1\leq k\leq N}(T_{2,k}/\omega - T_{1,k})$, with $i = \arg\min_{1\leq k\leq N}(T_{2,k}/\omega - T_{1,k})$, which does not correspond to the maximum d (i.e., not on the curve passing through point A in Figure 6.7). According to (6.34), ξ is minimized by choosing ω as small as possible. Taking into account the continuity of ξ with respect to ω and d, one can show that ξ is monotonically decreasing as long as ω is decreased until it reaches the value corresponding to the point A.

Now suppose that $\hat{d}$ lies on $\min_{1\leq k\leq N}(T_{4,k} - T_{3,k}/\omega)$ and let

$$j = \arg\min_{1\leq k\leq N}(T_{4,k} - T_{3,k}/\omega)$$

corresponding to the maximum d (i.e., point A in Figure 6.7), then ξ can be written as

$$\begin{aligned} \xi &= \sum_{k=1}^{N} \frac{T_{2,k} - T_{3,k}}{\omega} - \sum_{k=1}^{N} (T_{1,k} - T_{4,k}) - 2N\left(T_{4,j} - \frac{T_{3,j}}{\omega}\right) \\ &= \frac{1}{\omega}\sum_{k=1}^{N}\left(T_{2,k} - T_{3,k} + 2T_{3,j}\right) - \sum_{k=1}^{N}\left(T_{1,k} - T_{4,k} + 2T_{4,j}\right). \end{aligned} \tag{6.35}$$

From (6.35), it is clear that ξ can be minimized by taking the largest possible ω if the expression $\sum_{k=1}^{N}(T_{2,k} - T_{3,k} + 2T_{3,j})$ is positive and by taking the smallest possible ω if $\sum_{k=1}^{N}(T_{2,k} - T_{3,k} + 2T_{3,j})$ is negative as depicted by Figure 6.7. Hence, for

$$2NT_{3,j} > \sum_{k=1}^{N}(T_{3,k} - T_{2,k}),$$

the MLE is again given by (6.32) and (6.33), and for

$$2NT_{3,j} < \sum_{k=1}^{N}(T_{3,k} - T_{2,k}),$$

the MLE is given by the intersection of the curves $(T_{4,m} - T_{3,m}/\omega)$ and $(T_{4,n} - T_{3,n}/\omega)$ (denoting the intersections of the curves $d = T_{4,k} - T_{3,k}/\omega$ and $d = T_{4,l} - T_{3,l}/\omega$ as $d^{k,l}$, where $d^{k,l}$ satisfy the constraints (6.29) and (6.30)), with

$$(m,n) = \arg\max_{k,l}\left\{ d^{k,l} \mid 2NT_{3,k} < \sum_{r=1}^{N}(T_{3,r} - T_{2,r});\ 2NT_{3,l} > \sum_{r=1}^{N}(T_{3,r} - T_{2,r})\right\}. \tag{6.36}$$

Basically, the indices (m,n) in (6.36) identify the first vertex of the support region located to the left of the vertex A for which a change of sign occurs in

$2NT_{3,n} - \sum_{r=1}^{N} (T_{3,r} - T_{2,r})$. In Figure 6.7, this vertex is represented by the point B, and the MLE $(\hat{d}, \hat{\omega})$ in this case is given by

$$\hat{d} = T_{4,m} - \frac{T_{3,m}(T_{4,m} - T_{4,n})}{T_{3,m} - T_{3,n}}, \tag{6.37}$$

and

$$\hat{\omega} = \frac{T_{3,m} - T_{3,n}}{T_{4,m} - T_{4,n}}. \tag{6.38}$$

Lemma 6.3 *To the left of the point where* $\min_{1 \le k \le N} (T_{2,k}/\omega - T_{1,k})$ *and* $\min_{1 \le k \le N} (T_{4,k} - T_{3,k}/\omega)$ *intersect (i.e., point A in Figure 6.7), the boundary of the support region is formed by the curves* $(T_{4,k} - T_{3,k}/\omega)$, $k = 1, \ldots, N$, *in such a way that as* ω *increases, a curve* $(T_{4,m} - T_{3,m}/\omega)$ *forms the new boundary of the support region after intersecting the curve* $(T_{4,n} - T_{3,n}/\omega)$ *if and only if* $m < n$.

Proof. The curve $(T_{4,N} - T_{3,N}/\omega)$ assumes negative values for small ω and is monotonically increasing towards $T_{4,N}$ as ω increases. Similarly, the curve $(T_{4,1} - T_{3,1}/\omega)$ assumes negative value for small ω and asymptotically approaches $T_{4,1}$ as ω increases. All the curves $(T_{4,k} - T_{3,k}/\omega)$, $k = 1, \ldots, N$, are arranged in descending order for small ω and in ascending order for large ω and they intersect each other somewhere around the true value of ω. Since the slope of each curve $(T_{4,k} - T_{3,k}/\omega)$, $k = 1, \ldots, N$ is $T_{3,k}/\omega^2$, the slope of the curve with index m is less than the slope of the curve with index n if $m < n$. Therefore, as ω increases, a curve can form the new boundary of the support region by intersecting another curve only if its index is lower than the previous one.

Lemma 6.4 *The MLE* $(\hat{d}, \hat{\omega})$, *whether (6.32) and (6.33) or (6.37) and (6.38), is unique.*

Proof. Note that the likelihood function is continuous on the boundary of the support region because different curves intersect each other on the vertices due to which there are no jumps in ξ and consequently in the likelihood function. Now considering the fact that $2NT_{3,j} > \sum_{k=1}^{N} (T_{3,k} - T_{2,k})$ for $j = N$, let

$$q = \arg\max_{j} \{T_{3,j} \mid 2NT_{3,j} < \sum_{k=1}^{N} (T_{3,k} - T_{2,k})\}.$$

Then it must also be true that $2NT_{3,j} < \sum_{k=1}^{N} (T_{3,k} - T_{2,k})$ for all $j < q$, i.e., for $j = 1, \ldots, q-1$ and $2NT_{3,j} > \sum_{k=1}^{N} (T_{3,k} - T_{2,k})$ for all $j > q$, i.e., for $j = q+1, \ldots, N$. Figure 6.6 shows the sign of the term $2NT_{3,j} - \sum_{k=1}^{N} (T_{3,k} - T_{2,k})$ for each $j = 1, \ldots, N$. There will always be just one change, if any, in the sign of this term from positive to negative. Therefore, ξ can be minimized by making ω as large as possible on the curve $(T_{4,q+1} - T_{3,q+1}/\omega)$ and as small as possible on the curve $(T_{4,q} - T_{3,q}/\omega)$ (or on the curve $(T_{2,i}/\omega - T_{1,i})$ if there is no such q) as shown in Figure 6.6.

Algorithm 2: Finding $\hat{\omega}$ and $\hat{d}$ for d unknown, ϕ known

Find $d^{k,l} = \frac{T_{2,k}T_{4,l} - T_{1,k}T_{3,l}}{T_{2,k}+T_{3,l}}$; $\omega^{k,l} = \frac{T_{2,k}+T_{3,l}}{T_{1,k}+T_{4,l}}$; $\forall\ k = 1, \ldots, N$ and $\forall\ l = 1, \ldots, N$.
$(i,j) = \arg\min\limits_{k,l} \{d^{k,l}\}$;
if $2NT_{3,j} > \sum\limits_{k=1}^{N} (T_{3,k} - T_{2,k})$ **then**
  $\hat{d} = d^{i,j}$; $\hat{\omega} = \omega^{i,j}$;
else
  $k = j$;
  LABEL:
  Find $d^{k,l} = T_{4,k} - \frac{T_{3,k}(T_{4,k} - T_{4,l})}{T_{3,k} - T_{3,l}}$; $\omega^{k,l} = \frac{T_{3,k} - T_{3,l}}{T_{4,k} - T_{4,l}}$; $\forall\ l = 1, \ldots, N$.
  $m = k$; $n = \arg\max\limits_{l} \{\omega^{k,l} \mid \omega^{k,l} < \omega^{i,j}\}$;
  if $2NT_{3,n} > \sum\limits_{k=1}^{N} (T_{3,k} - T_{2,k})$ **then**
    $\hat{d} = d^{m,n}$; $\hat{\omega} = \omega^{m,n}$;
  else
    $k = n$;
    goto LABEL;
  end if
end if

This fact, combined with Lemma 6.3, proves that the intersection of the curves forming the MLE is always unique. It should be noted that under the most likely scenario, when node B is sending its timestamps to node A after short delays, the MLE is given by (6.32) and (6.33), but in the usually unlikely scenario of node B waiting a long period of time before sending one of its timestamps to node A, (6.37) and (6.38) can be the MLE only if $2NT_{3,j} < \sum_{k=1}^{N} (T_{3,k} - T_{2,k})$. Note that in this case, in addition to the previous N^2 intersections, $N-1$ additional intersections have to be compared for each j satisfying $2NT_{3,j} < \sum_{k=1}^{N} (T_{3,k} - T_{2,k})$. The whole procedure for finding this MLE is summarized in Algorithm 2. This algorithm proceeds in precisely the same steps as described above.

Now that we have obtained some insight into this problem for ϕ known, we next proceed with the situation when ϕ is unknown.

Case III: fixed delay d known, clock offset ϕ unknown

The likelihood function in this case is the same as (6.24), where d is fixed and known. The region over which the likelihood function is non-zero is given by the indicator

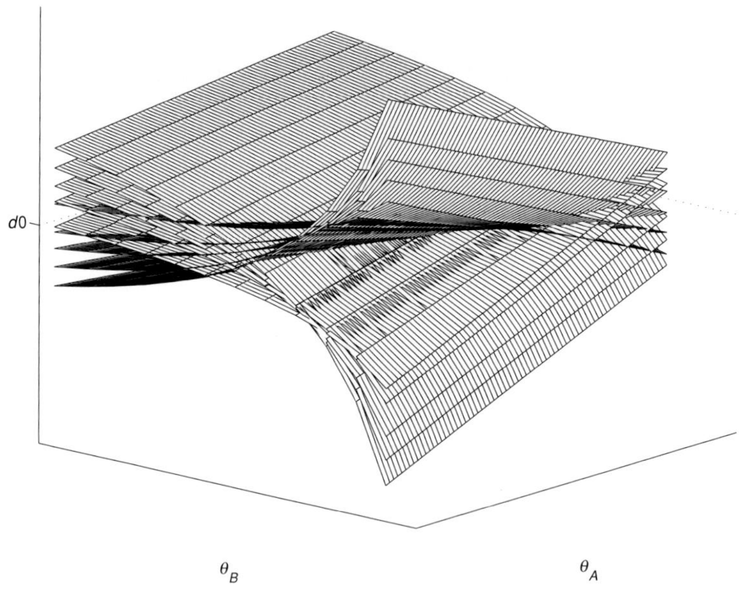

Figure 6.8: d as a function of ϕ and ω.

function $I[\cdot]$ in (6.24) and shown in Figure 6.8. This three-dimensional support region is dramatically more complex than what we observed in the first two cases. It is also evident from (6.24) that ξ is the same as in previous cases and the likelihood function can again be maximized by minimizing ξ. Since $\sum_{k=1}^{N}(T_{2,k} - T_{3,k})$ is always negative and d is given, ξ can be minimized by taking ω as small as possible. To find this minimum ω, we take a horizontal slice from this three-dimensional support region at the constant d. This gives an aerial view of the two-dimensional region shown in Figure 6.9, highlighting the relation between ω and ϕ for known d. Therefore, in accordance with (6.24), we can express the support of the likelihood function in the form of the following constraints:

$$-\infty \;<\; \phi \;<\; \infty, \tag{6.39}$$

$$\frac{T_{3,k} - \phi}{T_{4,k} - d} \;\leq\; \omega \;\leq\; \frac{T_{2,k} - \phi}{T_{1,k} + d}, \quad k = 1, \ldots, N. \tag{6.40}$$

These constraints can be viewed as ω being a monotonically decreasing function of ϕ for all k due to the positivity of $(T_{1,k} + d)$ and $(T_{4,k} - d)$, and the hatched region in Figure 6.9 is where these constraints are satisfied.

Lemma 6.5 *Of all the intersections of $(T_{2,k} - \phi)/(T_{1,k} + d)$ with $(T_{3,k} - \phi)/(T_{4,k} - d)$, only two points satisfy the constraints (6.39) and (6.40) in such a way that they represent*

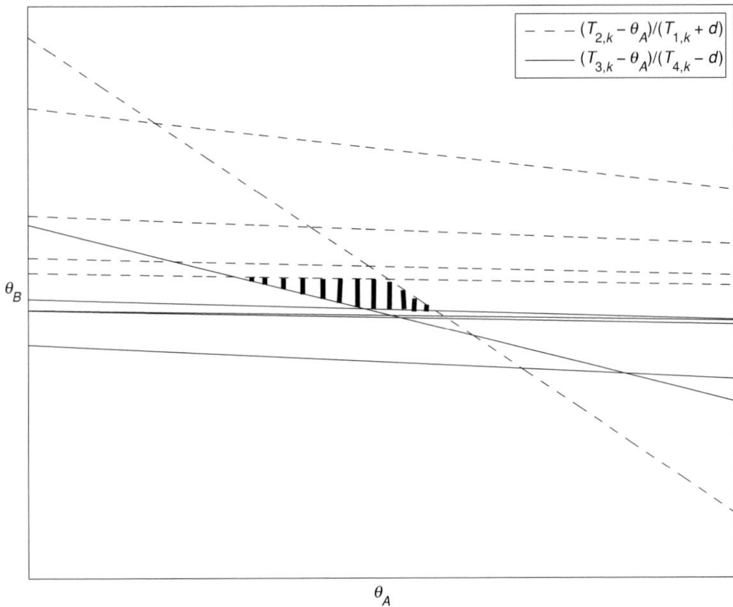

Figure 6.9: ω as a function of ϕ for constant d.

the starting and end points of the support region and the point with minimum ω is the one with maximum ϕ.

Proof. Consider the curves $(T_{2,k} - \psi)/(T_{1,k} + d)$ with $(T_{3,k} - \psi)/(T_{4,k} - d)$ as a function of ψ in order to avoid confusion between the actual unknown parameter ϕ and the variable with respect to which the above functions are drawn. Now utilizing (6.4) and (6.5), we can write

$$\frac{T_{2,k} - \psi}{T_{1,k} + d} = \left(1 + \frac{X_k}{T_{1,k} + d}\right)\omega + \frac{\phi - \psi}{T_{1,k} + d}, \qquad k = 1, \ldots, N,$$

$$\frac{T_{3,k} - \psi}{T_{4,k} - d} = \left(1 - \frac{Y_k}{T_{4,k} - d}\right)\omega + \frac{\phi - \psi}{T_{4,k} - d}, \qquad k = 1, \ldots, N.$$

It is clear that when $\psi = \phi$, $(T_{2,k} - \psi)/(T_{1,k} + d) > (T_{3,k} - \psi)/(T_{4,k} - d)$ for all k. Therefore, a support region *does* exist where the constraints (6.39) and (6.40) are satisfied. Now the slopes and y-intercepts of the straight lines $(T_{2,k} - \psi)/(T_{1,k} + d)$ are $-(T_{1,k} + d)^{-1}$ and $\left[1 + X_k(T_{1,k} + d)^{-1}\right]\omega + \phi(T_{1,k} + d)^{-1}$, respectively, and the slopes and y-intercepts of the straight lines $(T_{3,k} - \psi)/(T_{4,k} - d)$ are $-(T_{4,k} - d)^{-1}$ and $\left[1 - Y_k(T_{4,k} - d)^{-1}\right]\omega + \phi(T_{4,k} - d)^{-1}$, respectively. The y-intercepts can attain any value depending on the random delays X_k, Y_k, and the sign and magnitude of ϕ, but there is a set pattern in the slopes of these lines. According to the model (see Figure 6.1), it is always true that

$$(T_{1,1}+d)^{-1} > (T_{4,1}-d)^{-1} > (T_{1,2}+d)^{-1} > \cdots$$
$$> (T_{4,N-1}-d)^{-1} > (T_{1,N}+d)^{-1} > (T_{4,N}-d)^{-1}.$$

This is because $T_{1,1} < T_{4,1} < T_{1,2} < \cdots < T_{4,N-1} < T_{1,N} < T_{4,N}$. Due to the alternating slopes, the lines $(T_{2,k}-\psi)/(T_{1,k}+d)$ and $(T_{3,k}-\psi)/(T_{4,k}-d)$ for every k intersect at, at least, one point. According to the order of the slopes, both to the left and right of $\psi=\phi$, the support region ends after the first intersection. Therefore, there are exactly two points, $(\phi',\omega')_1$ and $(\phi',\omega')_2$, which define the starting and end point of the support region. In addition, the point corresponding to minimum ω' is the one with maximum ϕ' since all the straight lines always have negative slopes.

We can minimize ξ by taking the intersection of $\min_{1\leq k\leq N}(T_{2,k}-\phi)/(T_{1,k}+d)$ and $\max_{1\leq k\leq N}(T_{3,l}-\phi)/(T_{4,l}-d)$ at minimum possible $\hat{\omega}$, which gives the MLE $(\hat{\phi},\hat{\omega})$ as

$$\hat{\phi} = \frac{T_{3,j}(T_{1,i}+d) - T_{2,i}(T_{4,j}-d)}{(T_{1,i}+d)-(T_{4,j}-d)},$$
$$\hat{\omega} = \frac{T_{2,i}-T_{3,j}}{(T_{1,i}+d)-(T_{4,j}-d)},$$

where the indices (i,j) are the ones whose intersection gives the minimum allowed $\hat{\omega}$. Algorithm 3 presents in detail the steps that are required to find this MLE.

Algorithm 3 first finds all the intersections and chooses two candidate points $(\phi',\omega')_1$ and $(\phi',\omega')_2$ such that $\omega' \leq (T_{2,k}-\phi')/(T_{1,k}+d)$ for all k and $\omega' \geq (T_{3,l}-\phi')(T_{4,l}-d)$ for all l. These are the starting and end points of the non-zero likelihood region as proved in Lemma 6.5 and the point with minimum ω (which corresponds to the one with maximum ϕ) is chosen.

Algorithm 3: Finding $\hat{\phi}$ and $\hat{\omega}$ for d known, ϕ unknown

1. Find $\phi^{k,l} = \frac{T_{3,l}(T_{1,k}+d)-T_{2,k}(T_{4,l}-d)}{(T_{1,k}+d)-(T_{4,l}-d)}$; $\omega^{k,l} = \frac{T_{2,k}-T_{3,l}}{(T_{1,k}+d)-(T_{4,l}-d)}$; $\forall\ k=1,\dots,N$ and $\forall\ l=1,\dots,N$;
2. $(i,j) = \{(k,l)|\omega^{k,l} \leq \frac{T_{2,r}-\phi^{k,l}}{T_{1,r}+d}\ \forall\ r$ and $\omega^{k,l} \geq \frac{T_{3,r}-\phi^{k,l}}{T_{4,r}-d}\ \forall\ r\}$;
3. $(m,n) = \{(k,l)|(k,l)\neq(i,j),\ \omega^{k,l} \leq \frac{T_{2,r}-\phi^{k,l}}{T_{1,r}+d}\ \forall\ r$ and $\omega^{k,l} \geq \frac{T_{3,r}-\phi^{k,l}}{T_{4,r}-d}\ \forall\ r\}$;
4. $\hat{\omega} = \min\{\omega^{i,j},\omega^{m,n}\}$; $\hat{\phi} = \max\{\phi^{i,j},\phi^{m,n}\}$;

Case IV: fixed delay d unknown, clock offset ϕ unknown

In this case, all of d, ϕ, and ω are unknown and have to be jointly estimated. The likelihood function in this case is the same as in (6.24) but d is unknown. The region where the likelihood function is non-zero can be expressed in the form of the following constraints:

$$\begin{aligned} -\infty \ &< \ \phi \ < \ \infty, \ d \ > \ 0, \ \omega \ > \ 0, \\ d \ &\leq \ \frac{T_{2,k} - \phi}{\omega} - T_{1,k}, \quad k = 1, \ldots, N, \qquad (6.41) \\ d \ &\leq \ T_{4,k} - \frac{T_{3,k} - \phi}{\omega}, \quad k = 1, \ldots, N. \qquad (6.42) \end{aligned}$$

Within the constraint $d > 0$, $(T_{2,k} - \phi)/\omega - T_{1,k}$ are monotonically decreasing functions of ϕ and ω for all k, and $T_{4,k} - (T_{3,k} - \phi)/\omega$ are monotonically increasing functions of ϕ and ω for all k as shown in Figure 6.8. It is clear from the same figure that the non-zero likelihood region is similar in shape to a dome if we look at it standing on the (ϕ, ω) plane. Lemma 6.1 asserts that the MLE $(\hat{d}, \hat{\phi}, \hat{\omega})$ should lie somewhere on the ceiling of this dome. The lines on the (ϕ, ω) plane, on which the intersections of the surfaces lie, are given by

$$\phi = \frac{1}{2}\left[(T_{2,k} + T_{3,l}) - \omega(T_{1,k} + T_{4,l})\right], \quad k = 1, \ldots, N; \ l = 1, \ldots, N, \qquad (6.43)$$

or equivalently

$$\omega = \frac{T_{2,k} + T_{3,l} - 2\phi}{T_{1,k} + T_{4,l}}, \qquad k = 1, \ldots, N; \ l = 1, \ldots, N. \qquad (6.44)$$

Note that putting $\omega = 1$ (the case when there is no clock skew) in (6.43) and taking the minimum results in the MLE $\hat{\phi}$ in (6.25) derived in [41]. Although d is a function of both ϕ and ω, it can be written as a function of either ϕ only or ω only by utilizing this linear relationship between these two parameters. Figure 6.10 shows the *imaginary* two-dimensional region where d is drawn as a function of ω only, and Figure 6.11 shows the *imaginary* two-dimensional region where d is drawn as a function of ϕ only. Note that these are actually three-dimensional plots, but the points on the bottom two axes (ϕ, ω) are replaced with $(\frac{1}{2}\left[(T_{2,k} + T_{3,l}) - \omega(T_{1,k} + T_{4,l})\right], \omega)$ and $(\phi, (T_{2,k} + T_{3,l} - 2\phi)/(T_{1,k} + T_{4,l}))$ in Figures 6.10 and 6.11, respectively.

Over the line (6.43), d is given by

$$d \leq \frac{1}{2}\left[\frac{T_{2,k} - T_{3,l}}{\omega} + (T_{4,l} - T_{1,k})\right], \quad k = 1, \ldots, N; \ l = 1, \ldots, N. \qquad (6.45)$$

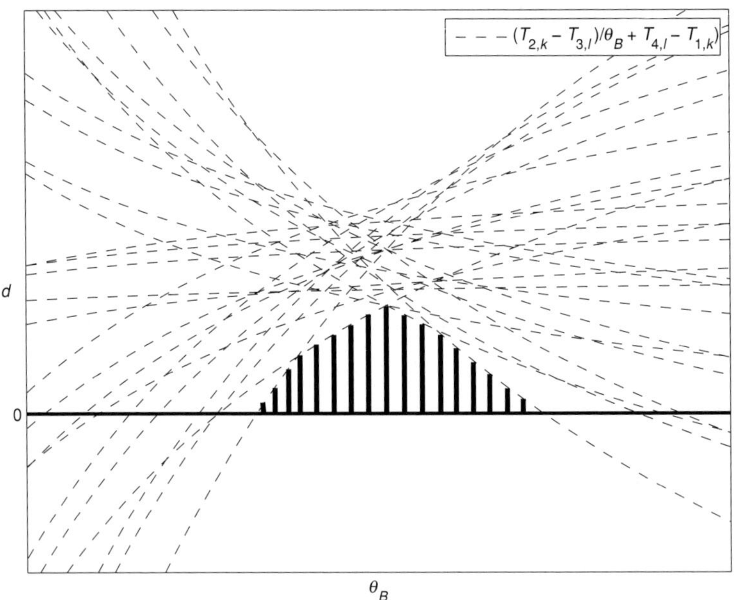

Figure 6.10: d as a function of ω only.

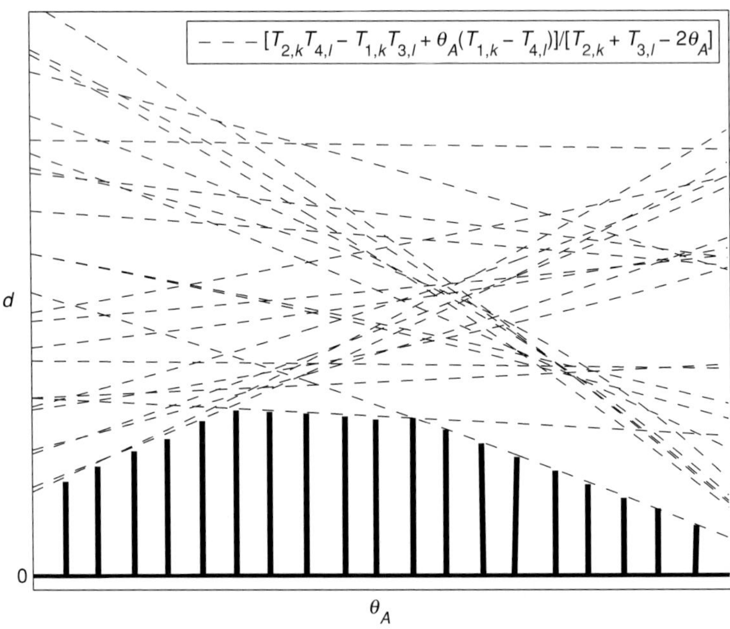

Figure 6.11: d as a function of ϕ only.

Note that putting $\omega = 1$ and taking the minimum results in the MLE $\hat{d}$ given in [41]. And over the line (6.44), d is given by

$$d \leq \frac{T_{2,k}T_{4,l} - T_{1,k}T_{3,l} + \phi(T_{1,k} - T_{4,l})}{T_{2,k} + T_{3,l} - 2\phi}, \quad k = 1, \ldots, N; \ l = 1, \ldots, N. \tag{6.46}$$

A closer look at (6.46) reveals that its right hand side converges to $-\infty$ or $+\infty$ respectively at $\phi = (T_{2,i} + T_{3,j})/2$ according to the negative or the positive sign of the numerator. But the constraint $d > 0$ automatically restricts the non-zero likelihood region well before even the first discontinuity of this kind as shown in Figure 6.11.

Estimating d and ω: Consider the set of N^2 curves given in (6.45) and plotted in Figure 6.10. Since the signs of $T_{2,k} - T_{3,l}$ and $T_{4,l} - T_{1,k}$ are always opposite, $N(N-1)/2$ for these curves have positive numerators in the term involving ω and negative constant terms, while the remaining $N(N+1)/2$ have negative numerators in the term involving ω and positive constant terms. Based on this observation, (6.45) can be written in the form of two sets of inequalities such that $(T_{2,k} - T_{3,l}) > 0$ for one set and $(T_{2,k} - T_{3,l}) < 0$ for the other as shown in Figure 6.10. Then the current scenario assumes a form quite similar to constraints (6.29) and (6.30). Therefore, initially a total of $[N(N-1)/2]\,[N(N+1)/2] = N^2(N^2-1)/4$ intersections (denoted by $d^{k,l,m,n}$ in Algorithm 4) are to be compared. Lemmas 6.1, 6.2, 6.3, and 6.4 are then similarly true for these sets of inequalities and the MLEs can be derived by following a similar procedure. Let us denote $\{\min_{1 \leq k,l \leq N} \frac{1}{2}\left[(T_{2,k} - T_{3,l})/\omega + (T_{4,l} - T_{1,k})\right] \mid (T_{2,k} - T_{3,l}) > 0\}$ as $(T_{2,i} - T_{3,j})/2\omega + (T_{4,j} - T_{1,i})/2$ and $\{\min_{1 \leq k,l \leq N} \frac{1}{2}\left[(T_{2,k} - T_{3,l})/\omega + (T_{4,l} - T_{1,k})\right] \mid (T_{2,k} - T_{3,l}) < 0\}$ as $(T_{2,m} - T_{3,n})/2\omega + (T_{4,n} - T_{1,m})/2$. Then if $\sum_{k=1}^{N}\left[T_{2,k} - T_{3,k} - (T_{2,m} - T_{3,n})\right]$ is positive, the MLE $(\hat{d}, \hat{\omega})$ is the intersection of this curve with the one discussed above, i.e.,

$$\hat{d} = \frac{1}{2}\left[\frac{(T_{2,i} - T_{3,j})((T_{1,i} - T_{1,m}) + (T_{4,n} - T_{4,j}))}{(T_{2,i} - T_{2,m}) + (T_{3,n} - T_{3,j})} + (T_{4,j} - T_{1,i})\right],$$

and

$$\hat{\omega} = \frac{(T_{2,i} - T_{2,m}) + (T_{3,n} - T_{3,j})}{(T_{1,i} - T_{1,m}) + (T_{4,n} - T_{4,j})}, \tag{6.47}$$

Otherwise, if $\sum_{k=1}^{N}\left[T_{2,k} - T_{3,k} - (T_{2,m} - T_{3,n})\right]$ is negative, then the MLE is the intersection of the curves $(T_{2,p} - T_{3,q})/2\omega + (T_{4,q} - T_{1,p})/2$ and $(T_{2,r} - T_{3,s})/2\omega + (T_{4,s} - T_{1,r})/2$ (denoting the intersections of the curves in (6.45) as $d^{k,l,m,n}$, for all

Algorithm 4: Finding $\hat{\phi}$, $\hat{\omega}$ and $\hat{d}$ for d unknown, ϕ unknown

$(m,n) = (1,N)$;
LABEL:
Find $d^{k,l,m,n} = \frac{1}{2}\left[\frac{(T_{2,k}-T_{3,l})((T_{1,k}-T_{1,m})+(T_{4,n}-T_{4,l}))}{(T_{2,k}-T_{2,m})+(T_{3,n}-T_{3,l})} + (T_{4,l}-T_{1,k})\right]$; $\omega^{k,l,m,n} = \frac{(T_{2,k}-T_{2,m})+(T_{3,n}-T_{3,l})}{(T_{1,k}-T_{1,m})+(T_{4,n}-T_{4,l})}$; $\forall\,(k,l) \neq (m,n)$;
$(p,q) = \arg\min_{k,l}\{d^{k,l,m,n}\}$
if $T_{2,p} - T_{3,q} > 0$ **then**
  $\hat{d} = d^{p,q,m,n}$; $\hat{\omega} = \omega^{p,q,m,n}$; $\hat{\phi} = \frac{1}{2}\left[(T_{2,p}+T_{3,q}) - \hat{\omega}(T_{1,p}+T_{4,q})\right]$;
else
  if $N(T_{2,p} - T_{3,q}) > \sum_{k=1}^{N}(T_{3,k} - T_{2,k})$ **then**
    $\hat{d} = d^{p,q,m,n}$; $\hat{\omega} = \omega^{p,q,m,n}$; $\hat{\phi} = \frac{1}{2}\left[(T_{2,p}+T_{3,q}) - \hat{\omega}(T_{1,p}+T_{4,q})\right]$;
  else
    Remove (m,n) curve;
    $(m,n) = (p,q)$;
    goto LABEL;
  end if
end if

(k,l,m,n), and $d^{k,l,m,n}$ satisfy the constraints (6.41) and (6.42)), where

$$(p,q,r,s) = \arg\max_{k,l,m,n}\left\{d^{k,l,m,n} \;\middle|\; N(T_{3,l}-T_{2,k}) < \sum_{k=1}^{N}(T_{3,k}-T_{2,k});\; N(T_{3,n}-T_{2,m}) > \sum_{k=1}^{N}(T_{3,k}-T_{2,k})\right\}.$$

Hence, here the MLE $(\hat{d},\hat{\omega})$ is given by

$$\hat{d} = \frac{1}{2}\left[\frac{(T_{2,p}-T_{3,q})((T_{1,p}-T_{1,r})+(T_{4,s}-T_{4,q}))}{(T_{2,p}-T_{2,r})+(T_{3,s}-T_{3,q})} + (T_{4,q}-T_{1,p})\right],$$

and

$$\hat{\omega} = \frac{(T_{2,p}-T_{2,r})+(T_{3,s}-T_{3,q})}{(T_{1,p}-T_{1,r})+(T_{4,s}-T_{4,q})}, \tag{6.48}$$

The complete procedure for finding the MLE is described in Algorithm 4. Although a modified Algorithm 2 can be used in this case, we present this alternative algorithm for the sake of completeness. It starts from the curve for which $(T_{2,m} - T_{3,n})$ is minimum, i.e., $(T_{2,1} - T_{3,N})$ and then compares its intersections with other curves. It keeps on replacing this curve with the one giving the next minimum $d^{k,l,m,n}$ within the constraints until the MLE is found according to the procedure described before.

Estimating ϕ: A simpler and easier to implement method is estimating $\hat{\phi}$ by noting that for every d as a function of ω (and hence the one minimizing ξ), there is a corresponding ϕ according to (6.43). Therefore, the MLE is

$$\hat{\phi} = \frac{1}{2}\left[(T_{2,i} + T_{3,j}) - \hat{\omega}(T_{1,i} + T_{4,j})\right], \tag{6.49}$$

or

$$\hat{\phi} = \frac{1}{2}\left[(T_{2,p} + T_{3,q}) - \hat{\omega}(T_{1,p} + T_{4,q})\right], \tag{6.50}$$

depending on whether $\hat{\omega}$ is given by (6.47) or (6.48). The reason for not following the same procedure as in finding $\hat{\omega}$ by using (6.44) is that the problem becomes computationally complex. First, the likelihood function assumes quite a complicated form after plugging (6.44) and (6.46) into (6.24). Second, the intersection $\hat{\phi}$ of the curves in (6.44) has to be found by solving quadratic equations with large coefficients. To be exact, $\hat{\phi}$ is the solution of

$$\begin{aligned}
&2\hat{\phi}^2[(T_{1,r} - T_{1,p}) + (T_{4,s} - T_{4,q})] + \hat{\phi}[(T_{1,p} - T_{4,q})(T_{2,r} + T_{3,s}) \\
&\quad - (T_{1,r} - T_{4,s})(T_{2,p} + T_{3,q})] + [T_{2,p}T_{2,r}(T_{4,q} - T_{4,s}) \\
&\quad + T_{2,p}T_{3,s}(T_{4,q} + T_{1,r}) - T_{2,r}T_{3,q}(T_{4,s} + T_{1,p}) - T_{3,q}T_{3,s}(T_{1,r} - T_{1,p})] = 0,
\end{aligned}$$

where the indices p, q, r, s are the ones minimizing ξ. This equation has two solutions and the solution which gives

$$\hat{\phi} < \min(T_{2,i} + T_{3,j})/2, \quad i = 1, \ldots, N;\ j = 1, \ldots, N,$$

is accepted to satisfy the constraints set by $I[\cdot]$ in (6.24). Hence, (6.49) or (6.50) should be chosen to estimate ϕ on the grounds of less computational complexity. It should be noted that $\hat{d}$ is the same in both approaches when we estimate it jointly with $\hat{\omega}$ and $\hat{\phi}$ whether by expressing it in terms of ω only or in terms of ϕ only. Algorithm 4 also includes the step for estimating ϕ. Computer simulations illustrating the performance of the proposed MLE under various modeling assumptions will be reported in Chapter 7.

Chapter 7

COMPUTATIONALLY SIMPLIFIED SCHEMES FOR ESTIMATION OF CLOCK OFFSET AND SKEW

Although the MLE derived in the previous chapter is not computationally very complex, WSNs can still benefit from some simplified schemes to estimate the clock parameters specially when the synchronization accuracy constraints are not extremely stringent but the energy conservation constraints are. In addition, to estimate both the clock offset and skew in the Gaussian noise case, knowledge of the fixed portions of delay d was required, which is not usually available beforehand. Therefore, in this chapter, two simple algorithms will be developed to estimate the clock offset and skew regardless of the distribution of the delays, and these are very suitable for the low-power-demanding regime of WSNs. The proposed estimators can be implemented using simple steps and present remarkably low complexity. These estimators and the derived performance bounds are targeting practical applications, and are of much significance due to their robustness to the actual distribution of network delays.

The main topics in this chapter are as follows. In the first proposed estimation scheme, the clock skew is estimated using only the first and the last data samples, since the difference between timestamps is largest between those two samples for any distribution, and then maximum-likelihood-like estimators (MLLEs) and Cramer–Rao-like lower bounds are derived for the clock skew. Subsequently, the data are processed to remove the effect of skew and then the clock offset is estimated, which just requires a few computations. The second proposed clock offset estimation scheme fits a line between two points, the differences between the first and the fourth timestamps, that are at a minimum distance apart, yielding both the clock offset and skew regardless of the underlying actual distribution.

7.1 Using the First and the Last Data Sample

Exploiting the fact that the clock difference between two wireless terminals is monotonically increasing (or temporarily decreasing then increasing) based on the linear clock skew model adopted in this chapter, the clock difference will be maximized between the first and last timestamps. From this idea, novel and practical clock skew estimators can be developed by using the first and last observations of timing message exchanges. Indeed, the proposed MLLE maximizes the likelihood function based on the reduced set of observations (the first and last timestamps).

Assume the signaling framework depicted by Figure 6.1. From (6.4), subtracting $T_{2,1}$ from $T_{2,N}$ leads to

$$T_{2,N} - T_{2,1} = (T_{1,N} - T_{1,1} + X_N - X_1)\,\omega. \tag{7.1}$$

Similarly from (6.5), subtracting $T_{4,1}$ from $T_{4,N}$ yields

$$T_{3,N} - T_{3,1} = (T_{4,N} - T_{4,1} + Y_1 - Y_N)\,\omega. \tag{7.2}$$

Define the differences for the first and last timestamps as $D_{(1)} \triangleq \sum_{k=2}^{N} D_{1,k} = T_{1,N} - T_{1,1}$, $D_{(2)} \triangleq \sum_{k=2}^{N} D_{2,k} = T_{2,N} - T_{2,1}$, $D_{(3)} \triangleq \sum_{k=2}^{N} D_{3,k} = T_{3,N} - T_{3,1}$, and $D_{(4)} \triangleq \sum_{k=2}^{N} D_{4,k} = T_{4,N} - T_{4,1}$, respectively, and $D_{j,k} \triangleq T_{j,k} - T_{j,k-1}$ for $j = 1, \ldots, 4$ and $k = 2, \ldots, N$. Then (7.1) and (7.2) can be rewritten, respectively, as

$$D_{(2)} = \left(D_{(1)} + P\right)\omega, \tag{7.3}$$

$$D_{(3)} = \left(D_{(4)} - R\right)\omega, \tag{7.4}$$

where $P \triangleq X_N - X_1$ and $R \triangleq Y_N - Y_1$. Next, we analyze the observation model described by (7.3) and (7.4) to derive the MLLE and the Cramer–Rao-like lower bounds for the clock skew.

7.1.1 Gaussian Delay Model

Since X_N, X_1, Y_N, and Y_1 are iid normally distributed RVs with variance σ^2, P and R are zero mean normally distributed RVs with variance $2\sigma^2$, respectively. Then the joint pdf of P and R is given by

$$f_{P,R}(p, r) = \frac{1}{4\pi\sigma^2} \exp\left[-\frac{1}{4\sigma^2}\left(p^2 + r^2\right)\right].$$

Hence, the likelihood function of the simplified observation model (7.3) and (7.4) takes the form

$$L\left(\omega', \sigma^2\right) = \frac{1}{4\pi\sigma^2} \exp\left\{-\frac{1}{4\sigma^2}\left[D_{(2)}^2\left(\omega' - \beta\right)^2 + D_{(3)}^2\left(\omega' - \gamma\right)^2\right]\right\},$$

where $\omega' \triangleq 1/\omega$, $\beta \triangleq D_{(1)}/D_{(2)}$ and $\gamma \triangleq D_{(4)}/D_{(3)}$. Differentiating the log-likelihood function with respect to ω' yields

$$\frac{\partial \ln L\left(\omega',\sigma^2\right)}{\partial \omega'} = -\frac{1}{2\sigma^2}\left[D_{(2)}^2\left(\omega'-\beta\right) + D_{(3)}^2\left(\omega'-\gamma\right)\right]. \tag{7.5}$$

Thus the proposed MLLE for the Gaussian delay model (GMLLE) is obtained by equating the log-likelihood function (7.5) to zero and assumes the expression

$$\hat{\omega}_{GMLLE} = \frac{1}{\hat{\omega}'} = \frac{D_{(2)}^2 + D_{(3)}^2}{D_{(1)}D_{(2)} + D_{(3)}D_{(4)}}. \tag{7.6}$$

Again, similar procedures can be applied to derive a lower bound for the GMLLE. The second-order derivative of the log-likelihood function becomes

$$\frac{\partial^2 \ln L\left(\omega',\sigma^2\right)}{\partial \omega'^2} = -\frac{D_{(2)}^2 + D_{(3)}^2}{2\sigma^2}. \tag{7.7}$$

The expected value of (7.7) is given by

$$E\left[\frac{\partial^2 \ln L\left(\omega',\sigma^2\right)}{\partial \omega'^2}\right] = -\frac{E\left[D_{(2)}^2 + D_{(3)}^2\right]}{2\sigma^2} = -\frac{D_{(1)}^2 + D_{(4)}^2 + 4\sigma^2}{2\sigma^2}.$$

Finally, the lower bound of the GMLLE is given by

$$\mathrm{var}(\hat{\omega}_{GMLLE}) \geq \frac{\left(\frac{\partial \omega}{\partial \omega'}\right)^2}{-E\left[\frac{\partial^2 \ln L(\omega',\sigma^2)}{\partial \omega'^2}\right]} = \frac{2\sigma^2\omega^2}{D_{(1)}^2 + D_{(4)}^2 + 4\sigma^2}. \tag{7.8}$$

Note that the complexity of the MLLEs is far less than that of the GMLE. In fact, for the GMLE, the numbers of required multiplications and additions are about $4N+6$ and $10N$, respectively, while both MLLEs require only a few multiplications and additions (less than five) regardless of the number of beacons N. Moreover, for the GMLE, the fixed portion of delays d must also be estimated, which requires additional computations.

7.1.2 EXPONENTIAL DELAY MODEL

For exponential delays, X_N, X_1, Y_N, and Y_1 are assumed to be iid exponentially distributed RVs with mean α. Therefore, P and R are zero mean Laplace distributed RVs with variance $2\alpha^2$. Thus, the joint pdf of P and R is given by

$$f_{P,R}\left(p,r\right) = \left(\frac{1}{2\alpha}\right)^2 \exp\left[-\frac{1}{\alpha}\left(|p|+|r|\right)\right].$$

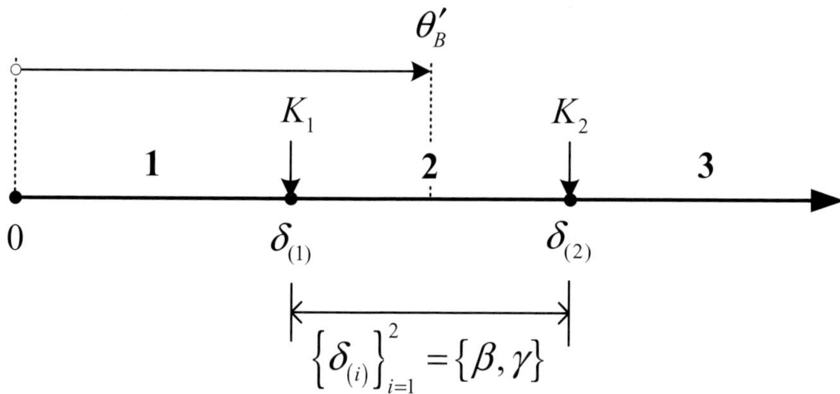

Figure 7.1: Regions of the order statistics $\{\delta_{(i)}\}_{i=1}^{2}$.

The likelihood function of the simplified observation model (7.3) and (7.4) is given by the expression

$$L(\omega, \alpha) = \left(\frac{1}{2\alpha}\right)^2 \exp\left[-\frac{1}{\alpha}\left(\left|\frac{D_{(2)}}{\omega} - D_{(1)}\right| + \left|D_{(4)} - \frac{D_{(3)}}{\omega}\right|\right)\right].$$

Substituting $1/\omega \triangleq \omega'$, the likelihood function can be rewritten as

$$L(\omega', \alpha) = \left(\frac{1}{2\alpha}\right)^2 \exp\left[-\frac{1}{\alpha}\left(D_{(2)}\left|\omega' - \beta\right| + D_{(3)}\left|\omega' - \gamma\right|\right)\right], \tag{7.9}$$

where $\beta \triangleq D_{(1)}/D_{(2)}$ and $\gamma \triangleq D_{(4)}/D_{(3)}$. The $\hat{\omega}'$ maximizing the likelihood function (7.9) is given by

$$\begin{aligned} \hat{\omega}' &= \arg\min_{\omega'}\left(D_{(2)}\left|\omega' - \beta\right| + D_{(3)}\left|\omega' - \gamma\right|\right), \\ &= \arg\min_{\omega'}\sum_{i=1}^{2} K_i\left|\omega' - \delta_{(i)}\right|, \end{aligned} \tag{7.10}$$

where the order statistics $\{\delta_{(i)}\}_{i=1}^{2}$ are generated from the given observations $\{\beta, \gamma\}$ and K_i is the distance either $D_{(2)}$ or $D_{(3)}$. Let $\hat{j} = \arg\min_j \sum_{i=1}^{2} K_i|\delta_{(j)} - \delta_{(i)}|$, then the proposed clock skew can be derived from the solution of the minimization problem (7.10), which can be written as

$$\hat{\omega}' = \arg\min_{\omega'}\sum_{i=1}^{2} K_i\left|\omega' - \delta_{(i)}\right| = \arg\min_{\omega'} h(\omega'),$$

where $h(\omega') \triangleq \sum_{i=1}^{2} K_i\left|\omega' - \delta_{(i)}\right|$. Now divide the region of order statistics $\{\delta_{(i)}\}_{i=1}^{2}$ into three different regions as in Figure 7.1, then the function $h(\omega')$ in the first region becomes

$$h(\omega') = -\sum_{i=1}^{2} K_i\omega' + \sum_{i=1}^{2} K_i\delta_{(i)} \qquad \omega' \le \delta_{(1)} \text{ (region } \mathbf{1}) .$$

Since K_i is always positive, the corresponding estimate $\hat{\omega}'$ is given by

$$\hat{\omega}' = \arg\min_{\omega'} h(\omega') = \delta_{(1)} \qquad \text{(region } \mathbf{1}) .$$

Similarly, in the second region, the function $h(\omega')$ takes the form

$$h(\omega') = (K_1 - K_2)\,\omega' + \left(K_2\delta_{(2)} - K_1\delta_{(1)}\right), \qquad \delta_{(1)} < \omega' \le \delta_{(2)} \text{ (region } \mathbf{2}) .$$

Hence the estimate $\hat{\omega}'$ is given by

$$\hat{\omega}' = \arg\min_{\omega'} h(\omega') = \begin{cases} \delta_{(1)} & K_1 > K_2 \\ \delta_{(2)} & K_1 < K_2 \\ \text{any value} & K_1 = K_2 \end{cases} \qquad \delta_{(1)} < \omega' \le \delta_{(2)} \text{ (region } \mathbf{2}) .$$

Finally, the function $h(\omega')$ in the final region is given by

$$h(\omega') = \sum_{i=1}^{2} K_i\omega' - \sum_{i=1}^{2} K_i\delta_{(i)}, \quad \delta_{(2)} < \omega' \text{ (region } \mathbf{3}) .$$

So the estimate $\hat{\omega}'$ in this region is given by

$$\hat{\omega}' = \arg\min_{\omega'} h(\omega') = \delta_{(2)} \qquad \text{(region } \mathbf{3}) .$$

Consequently, the estimate $\hat{\omega}'$ can be determined by choosing an appropriate value between the order statistics $\{\delta_{(i)}\}_{i=1}^{2}$. The median of $\{\delta_{(i)}\}_{i=1}^{2}$ maximizes the likelihood function and minimizes the MSE of the estimate. Therefore, the MLE of clock skew $\hat{\omega}$ for the exponential delay model is given by

$$\hat{\omega}_{EMLLE} = \begin{cases} \frac{D_2}{D_1}, & D_2 > D_3 \\ \frac{D_3}{D_4}, & D_2 < D_3 \\ \frac{1}{2}\left(\frac{D_2}{D_1} + \frac{D_4}{D_3}\right), & D_2 = D_3 \end{cases} . \tag{7.11}$$

Now we are interested in the lower bound of the exponential MLLE (EMLLE) to evaluate its asymptotic behavior. Despite the fact that the proposed likelihood function associated with the simplified model is not truly speaking an ML function, we will follow the same recipe as used in the derivation of CRLB to find a lower bound for the proposed EMLLE. The derivative of the log-likelihood function assumes the expression

$$\frac{\partial \ln L(\omega', \alpha)}{\partial \omega'} = \frac{D_{(2)}}{\alpha}\operatorname{sgn}(\omega' - \beta) + \frac{D_{(3)}}{\alpha}\operatorname{sgn}(\omega' - \gamma) . \tag{7.12}$$

Then the expected value of the square of (7.12) is given by

$$
\begin{aligned}
E\left[\left(\frac{\partial \ln L(\omega',\alpha)}{\partial \omega'}\right)^2\right] &= E_{P,R}\left[\frac{D_{(2)}^2 + D_{(3)}^2 + 2D_{(2)}D_{(3)}\mathrm{sgn}(\omega'-\beta)\,\mathrm{sgn}(\omega'-\gamma)}{\alpha^2}\right] \\
&\stackrel{(c)}{=} \frac{D_{(1)}^2 + D_{(4)}^2 + 4\alpha^2}{\alpha^2},
\end{aligned}
$$

where (c) is due to the fact that P and R are independent. Therefore, the lower bound of the EMLLE is given by

$$
\mathrm{var}(\hat{\omega}_{EMLLE}) \geq \frac{\left(\frac{\partial \omega}{\partial \omega'}\right)^2}{E\left[\left(\frac{\partial^2 \ln L(\omega',\alpha)}{\partial \omega'^2}\right)^2\right]} = \frac{\alpha^2\omega^2}{D_{(1)}^2 + D_{(4)}^2 + 4\alpha^2}. \tag{7.13}
$$

In fact, we have followed the same steps used in CRLB derivation since the same reasoning and proof can also be applied to the lower-bound derivation for the EMLLE.

7.1.3 Combination of Clock Offset and Skew Estimation

Since the proposed MLLEs are only for estimating clock skew ω, we still need to estimate clock offset ϕ for a complete clock synchronization. In the given clock skew model, $T_{2,k}$ and $T_{4,k}$ are known values and ω can be estimated using the MLLE, and the sets of delay observations between two nodes can be rewritten as

$$
U_k' = T_{2,k} - \hat{\omega} T_{1,k} \quad (= d' + \phi + X_k'), \tag{7.14}
$$

$$
V_k' = \hat{\omega} T_{4,k} - T_{3,k} \quad (= d' - \phi + Y_k'), \tag{7.15}
$$

where $X_k' = \omega X_k$, $Y_k' = \omega Y_k$, and $d' = \omega d$. Notice further to the signal model represented by (7.14) and (7.15) that the same clock offset estimators as in (6.2) and (6.18) for Gaussian and exponential delay models, respectively, could be applied. Thus, substituting the sets of delay observations gives the following clock offset estimators:

$$
\hat{\phi} = \frac{\min_{1\leq k\leq N} U_k' - \min_{1\leq k\leq N} V_k'}{2} \quad \text{(exponential delays)}, \tag{7.16}
$$

$$
\hat{\phi} = \frac{\overline{U_k'} - \overline{V_k'}}{2} \quad \text{(Gaussian delays)}. \tag{7.17}
$$

Consequently, the proposed joint clock offset and skew estimators consist of the following steps:

1. Estimate the clock skew using the proposed MLLE, either $\hat{\omega}_{EMLLE}$ or $\hat{\omega}_{GMLLE}$ according to the type of random delays.

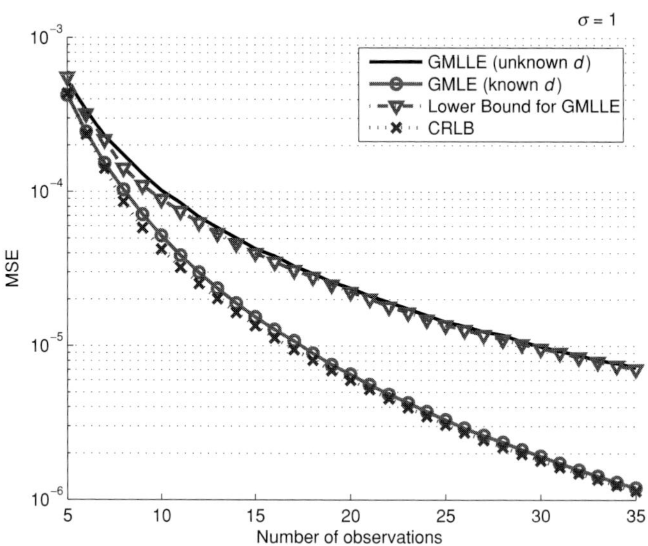

Figure 7.2: MSE of the MLE of the Gaussian delay model (GMLE) and the Gaussian MLLE (GMLLE) for Gaussian random delays ($\sigma = 1$).

2. Rewrite the sets of delay observations U'_k and V'_k as shown in (7.14) and (7.15).

3. Estimate the clock time offset using the estimator, (7.16) or (7.17), corresponding to the given delay model.

In fact, the proposed MLLEs require multiple message exchanges in a synchronization period ($N > 1$) to obtain the set of distances ($\{D_{(k)}\}_{k=1}^{4}$). However, these estimators can be applied not only within the same synchronization period, but also throughout several consecutive synchronization periods. In other words, a new set of observations in the next synchronization period can be substituted for the set of timestamps of the initial message exchange ($\{T_{k,N}\}_{k=1}^{4}$) in the initial synchronization period. This substitution can be sequentially performed thereafter. Therefore, the proposed MLLEs can also be applied to a single-message exchange model ($N = 1$) like TPSN without further modifications. The performance of the MLLEs is analyzed in the following section.

7.1.4 Simulation Results

Figure 7.2 compares the variance (MSE) of the GMLLE with the joint GMLE of clock skew and the corresponding CRLB when σ is 1. It can be seen that the performance of the GMLLE is close to that of the GMLE when the number of observations N is small (typically N is small in WSNs for the sake of energy efficiency), and

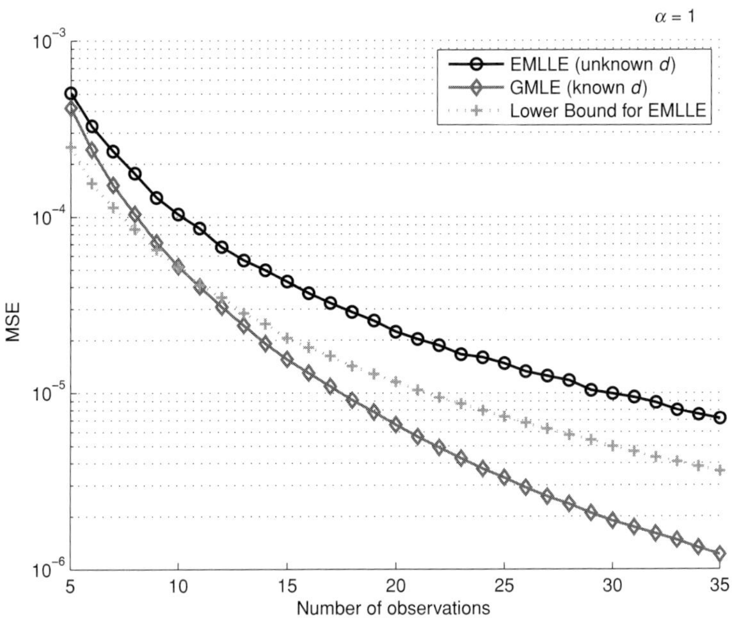

Figure 7.3: Variance of the GLME and the exponential MLLE (EMLLE) for exponential random delays ($\alpha = 1$).

its variance goes to zero as the number of observations increases (consistent and asymptotically efficient). Note that the GMLLE works well without the knowledge of the fixed portion of delays d, whereas this is required by the joint GMLE.

Figure 7.3 shows the variance of the EMLLE with the joint GMLE in exponential random delay channels when α is 1. It can be seen that again the proposed MLLE is consistent with and comparable to the GMLE. The consistency of the proposed MLLEs can be also checked from (7.13) and (7.8) since their lower bounds converge to zero as N increases.

In order to evaluate the robustness of estimators, in Figure 7.4 the performance of the GMLE is compared with that of the MLLEs in standard Gamma distributed (one of the most widely used distributions for modeling random queuing delays) random delay channels when γ is 2. Actually, both MLLEs exhibit similar performance compared to the GMLE regardless of the type of random delays. This is due to the fact that the performance of the MLLE is dominated by the set of distances ($\{D_{(k)}\}_{k=1}^{4}$), which does not vary much with respect to the type of random delays.

In Figure 7.5 the performance of the proposed clock offset estimator (7.17) is compared with that of the joint Gaussian MLE of clock offset derived in (6.11) in the Gaussian delay model when $\sigma = 0.5$. It can be seen that the joint MLE outperforms

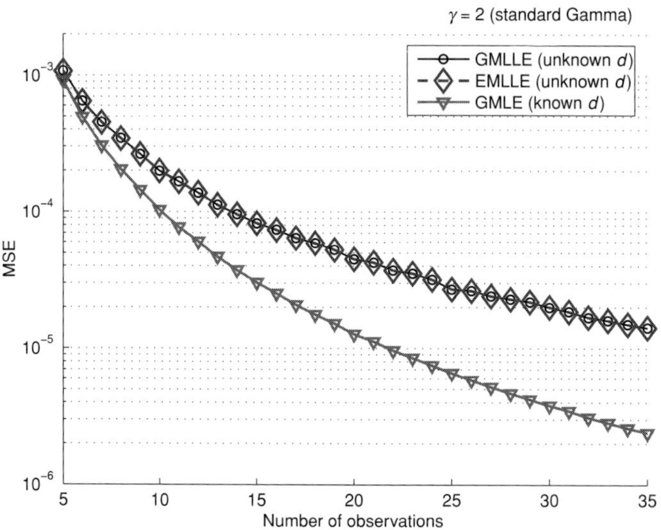

Figure 7.4: Variance of the GLME and the MLLEs for Gamma random delays ($\gamma = 2$).

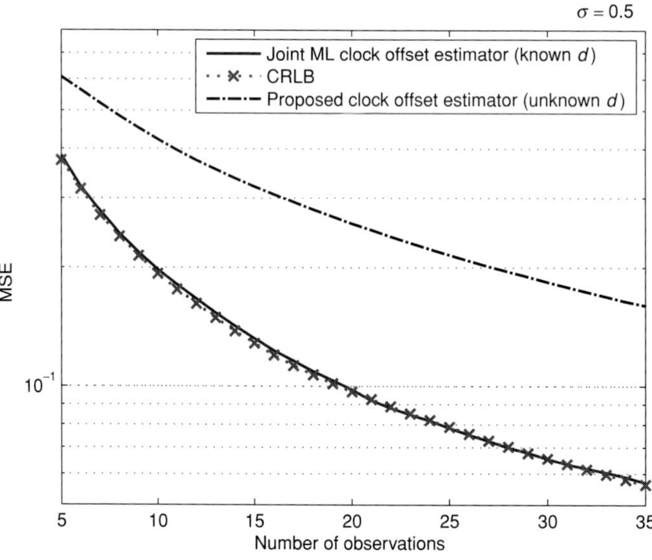

Figure 7.5: Variance of the joint ML clock offset estimate and the proposed estimator for Gaussian random delays ($\sigma = 0.5$).

the proposed estimator due to the help of the prior knowledge of d and the complete set of timestamps.

7.2 FITTING THE LINE BETWEEN TWO POINTS AT MINIMUM DISTANCE APART

In this section, we present an easier-to-implement algorithm which requires fewer computations at the expense of an increased MSE, and has the most desirable feature of independence with respect to the actual delay distribution incurred. The idea behind this algorithm is that (6.4) and (6.5) can be rewritten as

$$\begin{aligned} T_{2,k} &= T_{1,k}\omega + \phi + (d + X_k)\omega, \\ T_{3,k} &= T_{4,k}\omega + \phi - (d + Y_k)\omega. \end{aligned}$$

Notice that since ω, d, X_k, and Y_k are all positive, the points $T_{2,k}$, $k = 1, \ldots, N$ will always be above the line $T_{1,k}\omega + \phi$ and the points $T_{3,k}$, $k = 1, \ldots, N$ will always be below the line $T_{4,k}\omega + \phi$. Hence, a good estimate of ω and ϕ can be obtained by fitting a line between the observations such that $T_{2,k}$, $k = 1, \ldots, N$, are above the fitted line and $T_{3,k}$, $k = 1, \ldots, N$, are below it. The strategy that we follow for a good estimate is to join the two points P_1 and P_2, where P_1 corresponds to $\frac{1}{2}\min_{1\le k\le N}\{T_{4,k} - T_{1,k}\}$ and P_2 corresponds to $\frac{1}{2}\min_{1\le k\le N, k\neq i}\{T_{4,k} - T_{1,k}\}$. Representing their indices by i and j, respectively, we have

$$P_1 = \left\{\frac{1}{2}(T_{4,i} - T_{1,i}), \frac{1}{2}(T_{2,i} + T_{3,i})\right\},$$

and

$$P_2 = \left\{\frac{1}{2}(T_{4,j} - T_{1,j}), \frac{1}{2}(T_{2,j} + T_{3,j})\right\},$$

i.e., P_1 and P_2 correspond to the first two order statistics of the data set $\frac{1}{2}(T_{4,k} - T_{1,k})$, $k = 1, \ldots, N$. The line formed by joining those two points is shown in Figure 7.6 along with the true curve. Hence, the estimate $(\hat{\omega}, \hat{\phi})$ can be expressed as

$$\begin{aligned} \hat{\omega} &= \frac{(T_{2,i} + T_{3,i})/2 - (T_{2,j} + T_{3,j})/2}{(T_{1,i} + T_{4,i})/2 - (T_{1,j} + T_{4,j})/2}, \\ \hat{\phi} &= (T_{2,i} + T_{3,i})/2 - \hat{\omega}(T_{1,i} + T_{4,i})/2. \end{aligned}$$

When P_1 and P_2 fall very close to each other, the fitted line may go outside its boundaries and a part of it may become either greater than some $T_{2,k}$ or less than some $T_{3,k}$. In that case, we propose joining the minimum point P_1 with one of

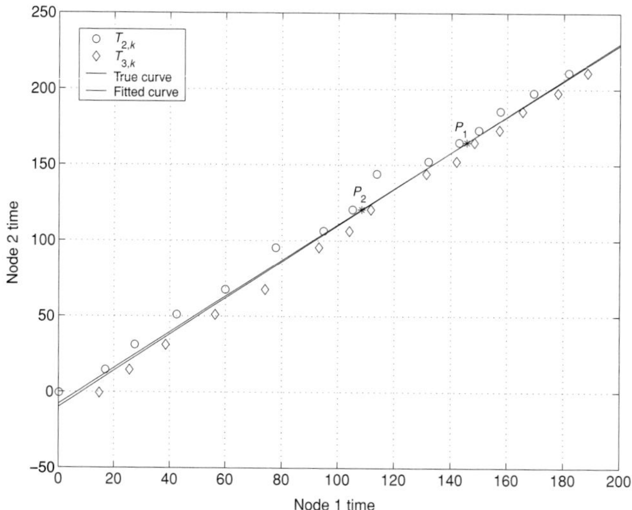

Figure 7.6: The estimated fit with the original curve.

the boundary points $\{T_{2,1}, T_{2,N}, T_{3,1}, T_{3,N}\}$ depending on which of them present the shortest distance from the initial fitted line. This algorithm is extremely simple since it just involves finding the first two order statistics from a set of N observations and checking the boundary conditions for the two extreme points. If the fitted line violates the boundary condition, the estimator is again formed by the same simple formula but with a point with a different time index. Since this point is on the boundary, the procedure does not have to be repeated and there are no loops involved as before. The whole procedure for finding these estimates is described in Algorithm 5. Some additional advantages of using Algorithm 5 are that ϕ can also be estimated by the y intercept of the fitted line and more importantly, d does not need to be known.

Algorithm 5: Fitting the line to estimate ω and ϕ

1. $i = \arg\min_{k} \frac{1}{2}\{T_{4,k} - T_{1,k}\}$;
2. $j = \arg\min_{k,\, k \neq i} \frac{1}{2}\{T_{4,k} - T_{1,k}\}$;
3. $\hat{\omega} = \frac{(T_{2,i}+T_{3,i})/2-(T_{2,j}+T_{3,j})/2}{(T_{1,i}+T_{4,i})/2-(T_{1,j}+T_{4,j})/2}$; $\hat{\phi} = (T_{2,i} + T_{3,i})/2 - \hat{\omega}(T_{1,i} + T_{4,i})/2$;
4. **if** $(T_{2,1} < \hat{\phi} + \hat{\omega}T_{1,1})$ or $(T_{2,N} < \hat{\phi} + \hat{\omega}T_{1,N})$ **then**
5. $\quad m = \arg\min_{k}\{|T_{2,k} - \hat{\phi} - \hat{\omega}T_{1,k}|, |T_{3,k} - \hat{\phi} - \hat{\omega}T_{4,k}|\}, \quad k = 1, N$;
6. $\quad \hat{\omega} = \frac{(T_{2,i}+T_{3,i})/2-(T_{2,m}+T_{3,m})/2}{(T_{1,i}+T_{4,i})/2-(T_{1,m}+T_{4,m})/2}$; $\hat{\phi} = (T_{2,i} + T_{3,i})/2 - \hat{\omega}(T_{1,i} + T_{4,i})/2$;
7. **end if**

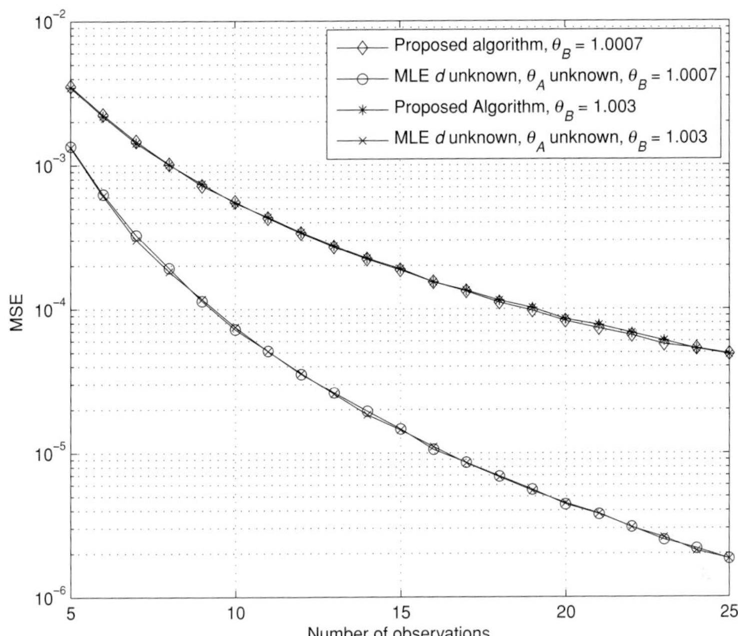

Figure 7.7: Comparison of the proposed algorithm with the MLE in Case IV.

7.2.1 SIMULATION RESULTS

As an example, we have simulated the performance of the MLE for fixed delay $d = 2$, clock offset $\phi = -10$, exponential delay parameter $\alpha = 2$, and for two different clock skews $\omega = 1.0007$ and $\omega = 1.003$. The reason for choosing different clock skews is to show a comparison of these algorithms in terms of performance for various parameters. We compare the performance of our proposed algorithm with the most general case when (d, ϕ, ω) have to be jointly estimated via the MLE in Case IV, as depicted in Chapter 6. In Figure 7.7 the MSEs of both clock skew estimators for $\omega = 1.0007$ and $\omega = 1.003$ is plotted against the number of message exchanges. It is clear from Figure 7.7 that although the MLE performs better than the proposed algorithm, it can still be adopted with the sacrifice of some performance degradation in the scenarios where the energy conservation is the main issue of concern. Hence, in the light of the accuracy–energy tradeoff for attaining such a gain in performance by deploying MLE, we assert that the proposed algorithm is very suitable for WSNs. Moreover, there is not any significant difference between the MSE of the MLE and that of the proposed algorithm for different sets of actual parameters, and hence, it is suited to the different types of sensor nodes used currently. To check the robustness of our proposed algorithm to possible model mismatches, we have plotted the performance of the MLE in the most general

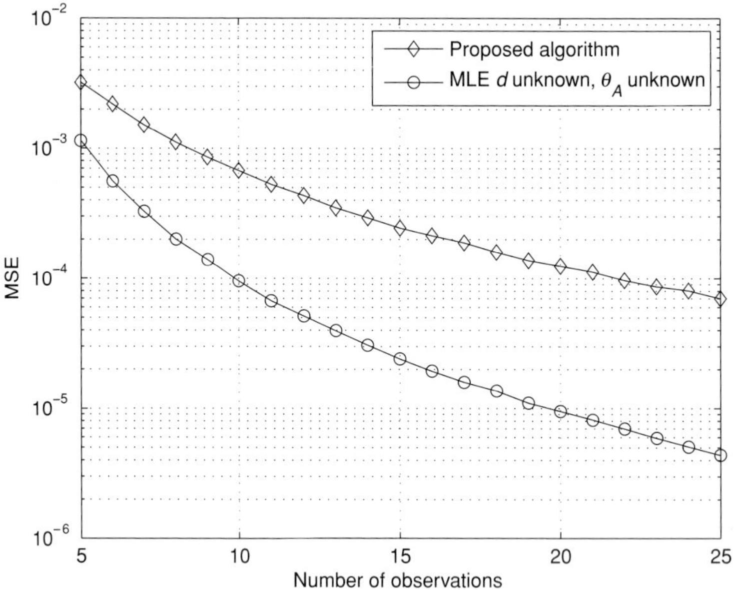

Figure 7.8: Comparison of the proposed algorithm with the MLE in Case IV for Gamma distributed random delays.

Case IV and our proposed algorithm in Figure 7.8 in which the actual random delays come from the widely used Gamma distribution instead of the exponential distribution. Figure 7.8 shows the MSE of both of these algorithms versus the number of observations when the random delays were simulated as Gamma RVs with shape parameter 2 and scale parameter 1. It is interesting to observe that the difference between their performances still remains on the same scale as in Figure 7.7. Therefore, the proposed algorithm is not only computationally simple and easy to implement but also robust to different environments.

7.2.2 COMPUTATIONAL COMPLEXITY COMPARISON

In Table 7.1 we present the number of operations required for the simplified algorithm in this section and the four algorithms used for deriving the MLE in Chapter 6. Note that these numbers have been calculated by considering the necessary simplifications (e.g., storing the output of an operation if it is to be used later). In addition, the operation count for Algorithm 2 and Algorithm 4 is given assuming no cycles. When their respective conditional statements become true, the code jumps around in the loop and the operation count is multiplied by the number of cycles. Moreover, it must be kept in mind that division is the most complex algorithm to implement in a digital signal processor and the number of division operations must be given the highest weight while choosing between different

Table 7.1: *Computational complexity of each algorithm*

	Additions	Multiplications	Divisions
Algorithm 1	$2N-1$	0	N
Algorithm 2 (1 cycle)	$5N^2+5N$	$2N^2+4$	$2N^2+2N$
Algorithm 3	$4N^3+5N^2+2$	$2N^2$	$2N^3+2N^2$
Algorithm 4 (1 cycle)	$9N^2+N+6$	$2N^2+5$	$2N^2$
Proposed algorithm	$3N+31$	$N+10$	2

algorithms. Finally, the operation count of our proposed algorithm is given for the worst case scenario, the probability of which is very low. For usual operation, its complexity will only be $3N+11$ additions, $N+4$ multiplications and 1 division.

For a comparison, observe that even for a small number of observations, e.g., 10, Algorithm 4 requires 916 additions, 205 multiplications, and 200 divisions. On the other hand, the proposed algorithm requires only 61 additions, 20 multiplications and 2 divisions for 10 observations in the worst case. As the number of observations N increases, the difference between the operation counts increases significantly while the difference in the MSE decreases, making it a more viable option for large N. However, it must be remembered that in the light of the results in [75], which reported that the energy required to transmit 1 bit over 100 meters (3 joules) is equivalent to the energy required to execute 3 *million* instructions, employing the MLE to achieve clock synchronization in a WSN is still a practical option.

Chapter 8

PAIRWISE BROADCAST SYNCHRONIZATION (PBS)

As discussed in Chapter 2, there are a number of key factors in designing time synchronization protocols for WSNs, such as accuracy, energy consumption, scalability, acquisition time, implementation complexity, and robustness. The most important and crucial factor is the tradeoff between accuracy and energy consumption. Increasing the synchronization accuracy in general requires more energy consumption to transmit the RF timing messages among sensor nodes. But, the energy consumption for synchronization should be kept as small as possible since the power resources of common wireless sensors are strictly limited and are not rechargeable in general. However, for most of the existing synchronization protocols, there is a lack of in-depth analysis to assess the energy-efficiency tradeoff of synchronization algorithms. This chapter describes in detail the characteristics of the PBS protocol which efficiently combines both SRS and ROS approaches (described in Chapter 4) to achieve network-wide synchronization with a significantly reduced number of synchronization messages, i.e., with less energy consumption.

The main topics in this chapter are as follows. First, there is a brief summary of the PBS technique used to achieve network-wide synchronization for single-cluster sensor networks based on ROS, the newly developed approach, described in Chapter 4. Second, the performance of PBS is analyzed and compared with those of other well-known protocols. Third, for the extension to general multi-cluster sensor networks, use of the network-wide pair selection algorithm and the group-wise pair selection algorithm is proposed to select the best synchronization sequence aiming at minimizing the overall energy consumption, respectively. Fourth, the performance of the proposed pair selection algorithms is analyzed with respect to the number of required synchronization messages (i.e., energy consumption).

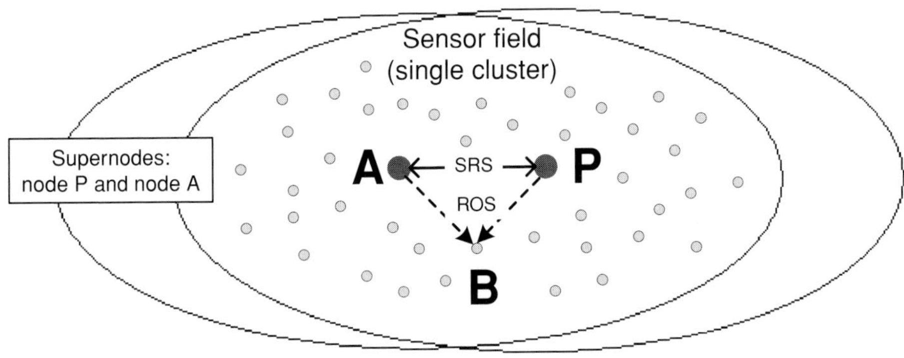

Figure 8.1: PBS for single-cluster networks.

8.1 SYNCHRONIZATION FOR SINGLE-CLUSTER NETWORKS

In Figure 8.1, every node in a single-cluster network (e.g., node B) can receive messages from both supernodes, node P and node A, while node P and node A perform a pairwise synchronization using two-way timing message exchanges as shown in Chapter 4. Therefore, the proposed PBS protocol achieves global synchronization by performing pairwise synchronization between the two supernodes using the ROS approach, and the joint clock offset and skew estimator for Gaussian random delays, derived in (4.12), can be used to reach pairwise synchronization.

8.2 COMPARISONS AND ANALYSIS

This section compares the proposed PBS protocol with other well-known synchronization protocols, such as TPSN, RBS, and FTSP, with respect to the energy consumption (number of required timing messages) and the synchronization accuracy.

Lemma 8.1 *Let N_{TPSN} be the required number of timing messages in TPSN, then $N_{TPSN} = 2N(L-1)$, where L stands for the number of overall sensor nodes in the network and N denotes the number of two-way message exchanges between two nodes.*

Proof. Since every node in the network, except a reference node, is connected to its parent node, there are $L-1$ branches (edges) in a hierarchical tree. Moreover, for TPSN, $2N$ timing messages are required in every pairwise synchronization. The number of required timing messages in TPSN is equal to the number of pairwise synchronizations times the number of required timing messages per pairwise synchronization, and therefore $N_{TPSN} = 2N(L-1)$.

Lemma 8.1 shows that N_{TPSN} is proportional to the number of nodes in the network. Note that this result can be applied to other level-based SRS protocols without loss of generality.

Lemma 8.2 *Let N_{FTSP} be the number of required timing messages in FTSP, then $N_{FTSP} = NL$.*

Proof. For FTSP, every sensor node must send its time readings upon receiving beacons (or broadcast beacons) to other nodes so that they can estimate the relative clock offsets between each other. Therefore, the number of required timing messages in FTSP is equal to the number of sensor nodes times the number of beacons: $N_{FTSP} = NL$. However, it must be noted that the one-way timing signaling mechanism is a more efficient scheme than two-way signaling, and the number of timing messages in FTSP is larger than in PBS only due to the protocol requirement that every node broadcasts its time for network flooding purposes. Otherwise, with respect to both the estimation efficiency and energy conservation, FTSP has the best performance of the synchronization protocols reported thus far.

Lemma 8.3 *Let N_{RBS} be the number of required timing messages in RBS, then $N_{RBS} = N + L(L-1)/2$.*

Proof. The reference node must broadcast the beacon packet N times in RBS. Moreover, every sensor node must send time readings upon receiving the broadcast beacons to all the other nodes in the network to compensate relative clock offsets between them [22]. Thus, $N_{RBS} = N + L(L-1)/2$, since the number of unique pairs in the network is $L(L-1)/2$.

When there are other sensor nodes which are located outside of the hatched region in Figure 4.1, the network could be divided into a number of separate groups (clusters) and they could be synchronized by additional pairwise synchronizations between supernodes in different groups, i.e., global synchronization can be achieved by a sequence of pairwise synchronizations. Here, diverse grouping and pair selection algorithms can be considered according to the type of the network. For instance, assuming that the level hierarchy of the network is discovered by an appropriate searching algorithm (e.g., as in [29]), there exist groups of parent and child nodes, where a group consists of a parent and its child nodes. Here, every parent node can investigate the connectivity between its child nodes and select the best sequence of synchronization pairs in order to minimize the required number of pairwise synchronizations, which maximizes the number of nodes performing ROS. Note that no network-wide heuristic connectivity search is required in this case because of its limited and known set of scanning nodes. The detailed extension of these preliminary considerations for the proposed PBS scheme is presented in the following sections.

The synchronization accuracy is another crucial design factor. In general, it depends on a variety of different factors, such as the network platform and setup, channel status, and estimation schemes. The performance of existing protocols has been compared in terms of the synchronization accuracy in various references, e.g., [2], [58], [84], and [94]. As shown in Chapter 4, the accuracy of PBS is exactly the same as that of RBS.

When there are multiple synchronization clusters (groups) in the network, the proposed PBS requires a series of pairwise synchronizations to achieve network-wide synchronization, i.e., an independent pairwise synchronization is necessary in every level of the network. Hence, the performance of ROS at each level (depth) of the network is independent of that of ROS at another level. Using the result that the MLE asymptotically approaches a Gaussian distribution irrespective of the distribution of the noise [50], the clock estimation error is also Gaussian. Consequently, the cumulative network-wide synchronization error can be modeled as a sum of normal RVs, i.e., as another normal RV. From (4.14), the variance of the network-wide synchronization error can be approximated as

$$\mathrm{var}(\hat{\phi}) \approx \sum_{i=1}^{d_{max}} \mathrm{var}(\hat{\phi}^{(i)}) \geq \sum_{i=1}^{d_{max}} \mathrm{CRLB}(\hat{\phi}^{(i)}), \tag{8.1}$$

where $\mathrm{CRLB}(\hat{\phi}^{(i)})$ denotes the CRLB for the clock offset estimator at the ith level of the network and d_{max} denotes the maximum depth (level) of the network.

To achieve a network-wide global synchronization, the following crucial question should be answered: How is the optimum set of pairwise synchronizations to minimize the number of timing message exchanges to be selected? The next sections answer this question and show an approach that guarantees network-wide synchronization.

8.3 SYNCHRONIZATION FOR MULTI-CLUSTER NETWORKS

There are two general approaches for extending the proposed PBS to general multi-cluster networks. When there is no problem with the placement of supernodes in the right positions of the network, the whole sensor field can be divided into several clusters, where each cluster contains two supernodes whose communication ranges cover the entire cluster. Hence, every cluster can be first synchronized by performing a pairwise synchronization between a pair of supernodes. Then, like in RBS, the global synchronization can be achieved by additional message exchanges (based on SRS) between supernodes in different clusters. In this case, the extension of PBS becomes mostly a problem of network implementation just like the cell-planning problem in mobile communication networks.

However, if either deploying supernodes or deploying them where we want them to be placed is not possible, there is no way to apply the above mentioned procedures. For this general scenario, an energy-efficient pair selection algorithm, called the Group-wise Pair selection Algorithm (GPA), is proposed later in the chapter to achieve global synchronization using ROS. First, we show a way to achieve global synchronization based on the network-wide heuristic search in order to reveal some general ideas of the pair selection problem as a preliminary study. Then, the proposed GPA is presented in detail.

8.3.1 NETWORK-WIDE PAIR SELECTION ALGORITHM (NPA)

Considering the energy efficiency in time synchronization, the problem of finding the optimum set of pairwise synchronizations is equivalent to that of minimizing the number of overall pairwise synchronizations in the network. There are two fundamental criteria for selecting the best synchronization pairs. First, a pair of nodes with the maximum number of nodes in their common coverage region of the pairwise synchronization has to be chosen during each selection step of the synchronization pair. Second, the depth (level) of the pairwise synchronization, i.e., the number of required successive pairwise synchronizations that is necessary to reach the reference node, has to be minimized since in general the synchronization errors increase with the depth of the network as shown in (8.1). To find the best pair, information about the network hierarchy and connectivity, which can be obtained by beacon exchanges between nodes, is required. Notice further that the network connection hierarchy can be constructed by applying the well-known breath-first search algorithm [8]. Here, every node in the network is required to send messages with their maximum power level satisfying a certain energy constraint.

As a graphical illustration of the proposed algorithms, an example of a network connection hierarchy is shown in Figure 8.2. Pairwise synchronization (PS) begins with the reference node, node 1, and four different branches (edges) are connected to the reference, i.e., there are four different nodes which can be chosen for the first synchronization pair. As mentioned, the criterion of selecting the best pair is to find a pair of nodes maximizing the number of synchronizing nodes (based on the ROS approach) from the pairwise synchronization. Let $p_{i,j}$ denote the pairwise synchronization between node i and node j, and $\mathbf{p}$ represent the pairwise synchronization sequence vector whose elements are a set of $p_{i,j}$. Denote by $N_{ROS}^{i,j}$ the number of synchronizing nodes, which are performing ROS from $p_{i,j}$. In Figure 8.2, node 4 must be selected for the first pair node since $N_{ROS}^{1,4} = 3$, which is the maximum achievable value of all possible choices (all the other nodes in level 1, nodes 2, 3, and 5, can be synchronized from $p_{1,4}$). The same criterion can be applied to determine the next pair of nodes thereafter, until all the nodes in the network are synchronized. Therefore, $p_{3,8}$, $p_{4,11}$, and $p_{11,14}$ are chosen as the second, third, and fourth pairs,

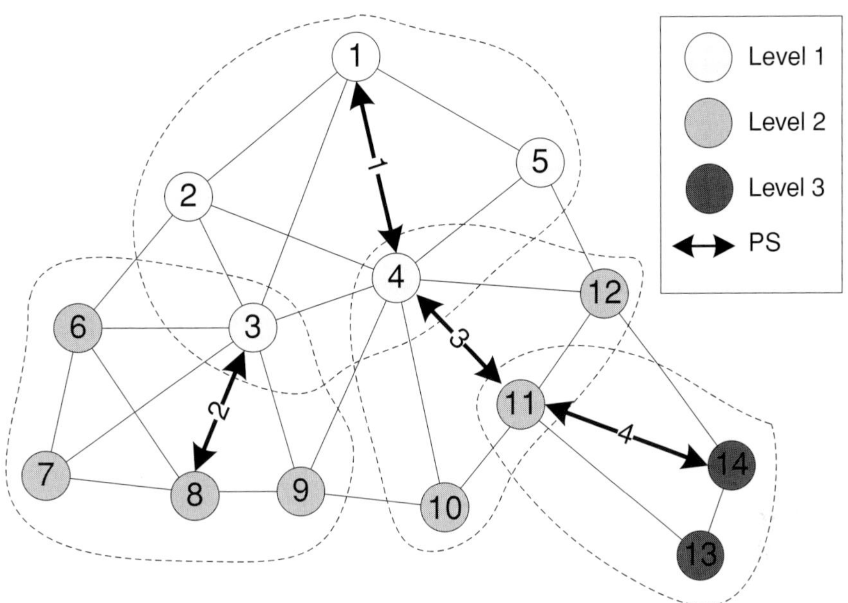

Figure 8.2: Network connection hierarchy.

respectively. Thus, a sequence of pairwise synchronizations is chosen to maximize the number of nodes performing ROS. In this example, the pairwise synchronization sequence vector is given by $\mathbf{p} = \{p_{1,4}, p_{3,8}, p_{4,11}, p_{11,14}\}$.

For the given pair selection criteria, we present next the NPA to find a pairwise synchronization sequence as a preliminary example. The network can be represented as a graph of $G = (V, E)$, where V represents the set of nodes (vertices) and E stands for the set of edges (branches), whose elements are two-element subsets of V, e.g., in Figure 8.2, $V = \{s_i\}_{i=1}^{14}$. Let L_i denote the subset of nodes located on level (depth) i, then $L_0 = \{s_1\}$, $L_1 = \{s_i\}_{i=2}^{5}$, $L_2 = \{s_i\}_{i=6}^{12}$, and $L_3 = \{s_{13}, s_{14}\}$ for the example depicted by Figure 8.2. Also, let S denote a set of synchronized nodes whose initial element is $S = \{s_1\}$, and $M_{i,j}$ denote the ith row and jth column element of the adjacency matrix M of the graph G, where $M_{i,j} = 1$ when node i and node j are connected, and $M_{i,j} = 0$ otherwise.

Note that an arbitrary node node k can be synchronized from $p_{i,j}$ if, and only if, nodes i and j are connected and node k is connected to both nodes i and j, i. e., $M_{i,j} = M_{i,k} = M_{j,k} = 1$. Moreover, the level of the nodes in a synchronization pair must differ by one level. Therefore, the number of synchronizing nodes from $p_{1,i}$ ($N_{ROS}^{1,i}$) is given by

$$N_{ROS}^{1,i} = \sum_{j \neq i} M_{1,i} M_{1,j} M_{i,j}, \qquad \forall\, s_i \notin S,\ s_j \notin S,$$

where $s_i \in L_1$ and $s_j \in L_1$. Hence, the first pair node can be obtained by maximizing $N_{ROS}^{1,i}$:

$$\hat{\imath} = \arg\max_i N_{ROS}^{1,i},$$

when $s_i \in L_1$, otherwise no connection exists between node 1 and node i. In the example in Figure 8.2, $\hat{\imath} = 4$ because $N_{ROS}^{1,i}$ is 3 and achieves the maximum value. Thus, $p_{1,4}$ is selected as the first pair. Note that a pair of nodes in the same level should not be selected as a valid pair in order to limit the bound for the maximum synchronization error which is proportional to the depth of the pairwise synchronization tree according to the second selection criterion. Hence, in general, to find the second pair of nodes, another node in L_1 should be chosen until all the nodes in L_1 are synchronized. However, in this example, there are no remaining unsynchronized nodes in L_1 after $p_{1,4}$ since all the nodes in L_1 are already synchronized by $p_{1,4}$ ($S = \{L_0, L_1\}$).

The same maximization procedure can be applied to find the next synchronization pair. A general formula for finding $N_{ROS}^{i,j}$ is given by

$$N_{ROS}^{i,j} = \sum_{k \neq j} M_{i,j} M_{i,k} M_{j,k} \qquad \forall\, s_i \in S,\ s_j \notin S,\ s_k \notin S, \tag{8.2}$$

where s_i is a candidate for the next parent node and the levels of s_j and s_k are equal and a level higher than that of the parent node. Again, based on the selection criterion, a node in a lower level has priority to be chosen as a parent node of the next pairwise synchronization. Likewise, the next synchronization pair can be found by maximizing $N_{ROS}^{i,j}$:

$$(\hat{\imath}, \hat{\jmath}) = \arg\max_{i,j} N_{ROS}^{i,j}. \tag{8.3}$$

Here, $p_{\hat{\imath},\hat{\jmath}}$ becomes the next element of $\mathbf{p}$ and all synchronized nodes from $p_{\hat{\imath},\hat{\jmath}}$ are added to S. From (8.2) and (8.3), the second synchronization pair becomes $p_{3,8}$ in this example since $N_{ROS}^{3,8}$ is 4 and maximum among all possible combinations of i and j. Thus, $\mathbf{p}$ becomes $\{p_{1,4}, p_{3,8}\}$ and $S = \{L_0, L_1, \{s_i\}_{i=6}^{9}\}$. Likewise, the third pair is chosen to be $p_{4,11}$, $\mathbf{p} = \{p_{1,4}, p_{3,8}, p_{4,11}\}$, and $S = \{L_0, L_1, L_2\}$. Repeating the same procedure (here, $s_i \in L_2$) gives $p_{11,14}$ as the last synchronization pair, and hence a complete sequence becomes $\mathbf{p} = \{p_{1,4}, p_{3,8}, p_{4,11}, p_{11,14}\}$ as depicted in Figure 8.2. Figure 8.3 illustrates the NPA.

8.3.2 GROUP-WISE PAIR SELECTION ALGORITHM (GPA)

To discover the overall network connectivity, every single node in the network has to transmit the connection discovery beacons and send back acknowledgement packets upon receiving other beacons from its adjacent nodes (e.g., the breath-first search

NETWORK-WIDE PAIR SELECTION ALGORITHM
Input: Graph (G), Adjacency matrix (M),
Maximum level/depth (d_{max})
Output: PS sequence vector ($\mathbf{p}$)
Initial values: $n = m = 1$

1 **while** $n \leq d_{max} - 1$
2 **for all** i, j, and k
($s_i \in S$, $s_i \in L_{n-1}$, and $s_j \notin S, s_k \notin S$, $s_j \in L_n, s_k \in L_n$)
3 $N_{ROS}^{i,j} \leftarrow \sum_{k \neq j} M_{i,j} M_{i,k} M_{j,k}$
4 $(\hat{i}, \hat{j}) \leftarrow \arg\max_{i,j} N_{ROS}^{i,j}$.
5 $\mathbf{p}(m) \leftarrow p_{\hat{i},\hat{j}}$
6 $m \leftarrow m + 1$
7 **If** any j, $s_j \in L_n$ and $s_j \notin S$, exists
8 **then** repeat from 2 to 6
9 **else** $n \leftarrow n + 1$

$*$ $\mathbf{p}(m)$: mth element of $\mathbf{p}$

Figure 8.3: Network-wide pair selection algorithm (NPA).

algorithm in [8]). For WSNs consisting of a large number of nodes, discovering the network connectivity is not a simple task and in general requires a number of packet exchanges. Therefore, instead we propose an efficient alternative method, the GPA, which relies on the hierarchical structure (spanning tree) of the network to simplify the connection discovery procedure in NPA. Note that the hierarchical tree of the network can be generated by a level discovery procedure as discussed in [29]. Once a hierarchical tree is established, there exist groups of parent and child nodes, where a group consists of a parent and its child nodes. In GPA, instead of discovering the entire network connectivity, every parent node only investigates the connectivity among its child nodes. Therefore, the reference node does not need to find the pairwise synchronization sequence of the entire network, but needs only to find the pairwise synchronization sequence among its child (level 1) nodes, and the other parent nodes successively perform the same connection searching procedure as the reference node. As a result, GPA significantly reduces the complexity of building a connection hierarchy, and requires a much smaller number of connection discovery beacons than NPA due to its limited and known set of scanning nodes. Furthermore, the work loads to find the best pairwise synchronization sequence can be balanced by sharing the roles of connection discovery and pair selection with the

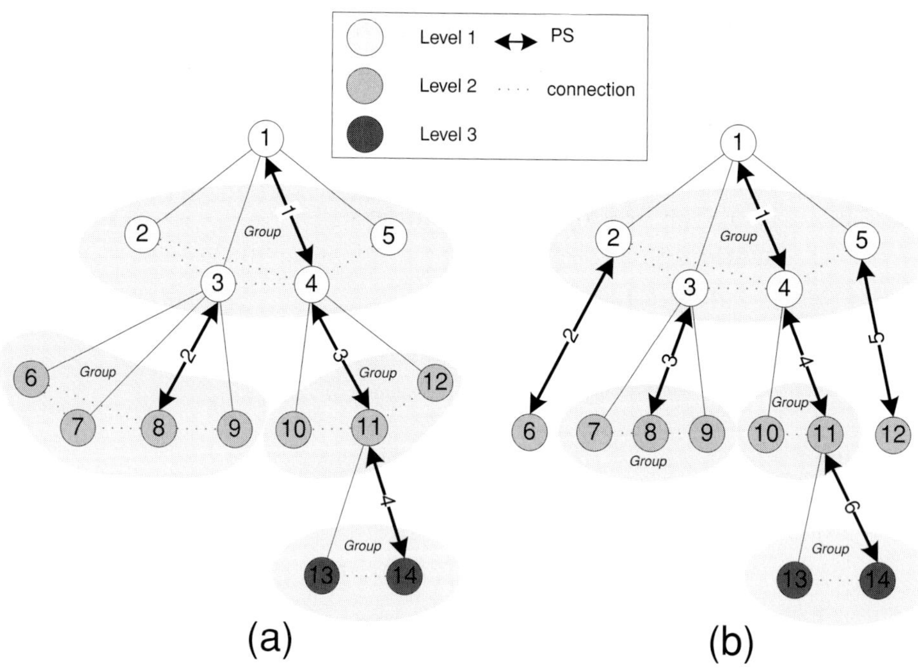

Figure 8.4: Examples of different hierarchical spanning trees of the network: (a) with four pairwise synchronizations; (b) with six pairwise synchronizations.

reference node and other parent nodes, i.e., no network-wide heuristic connection search is required for GPA. Figure 8.4 shows some possible hierarchical trees of the sample network. In Figure 8.4(a), the network can be synchronized using GPA with the same number of pairwise synchronizations as NPA, and nodes 4, 8, 11, and 14 are chosen as parent nodes. However, the number of pairwise synchronizations for GPA depends on the specific hierarchical tree, which is randomly constructed, and is greater than NPA in general. For instance, for another possible tree of the network as in Figure 8.4(b), the required number of pairwise synchronizations is six instead of four. Note that, for the same example, TPSN requires 13 pairwise synchronizations, the same as the overall number of branches (edges). The proposed GPA is presented in Figure 8.5, and the connection discovery process for GPA is summarized in the next subsection.

Group-Wise Connection Discovery

As the level discovery phase in TPSN [29], GPA first creates a hierarchical structure (spanning tree) of the network, then it searches the connection status among a set of child nodes in every parent–children group. The connection discovery procedure in GPA consists of the following steps:

GROUP-WISE PAIR SELECTION ALGORITHM
Input: Graph (G) and adjacency matrix (M) of each group
Output: PS sequence vector ($\mathbf{p}$) of each group
Initial value: $n = 1$

1 **for each** group whose parent is s_i
2   **for all** j and k ($s_j \notin S, s_k \notin S$, and children of s_i)
3     $N_{ROS}^{i,j} \leftarrow \sum_{k \neq j} M_{j,k}$
4   $\hat{j} \leftarrow \arg\max_{j} N_{ROS}^{i,j}.$
5   $\mathbf{p}(n) \leftarrow p_{i,\hat{j}}$
6   $n \leftarrow n + 1$
7   **If** any j, $s_j \notin S$, exists
8   **then** repeat from 2 to 6

$*$ $\mathbf{p}(n)$: nth element of $\mathbf{p}$

Figure 8.5: Group-wise pair selection algorithm (GPA).

1. A reference node is selected using an appropriate leader election algorithm (or pick a node having the highest priority) and assign it to a zero level.

2. A level discovery packet containing the identity and the level of packet is broadcast.

3. Every node that receives a level discovery packet is assigned a level (depth) that is one higher than that of the received packet and sends a new level discovery packet attaching its own level.

4. This process is repeated until every node in the network successfully is assigned a level.

5. Once a hierarchical tree is established, every parent–children group performs the following operation: every child node broadcasts a connection discovery packet to the other child nodes and sends back acknowledgement packets upon receiving other connection discovery packets.

After being assigned a level, every node discards further packets requesting level discovery to prevent collisions. In addition, connection discovery packets from any child node belonging to other groups will also be discarded. Notice also that other algorithms can be considered when constructing the spanning tree [32], [99].

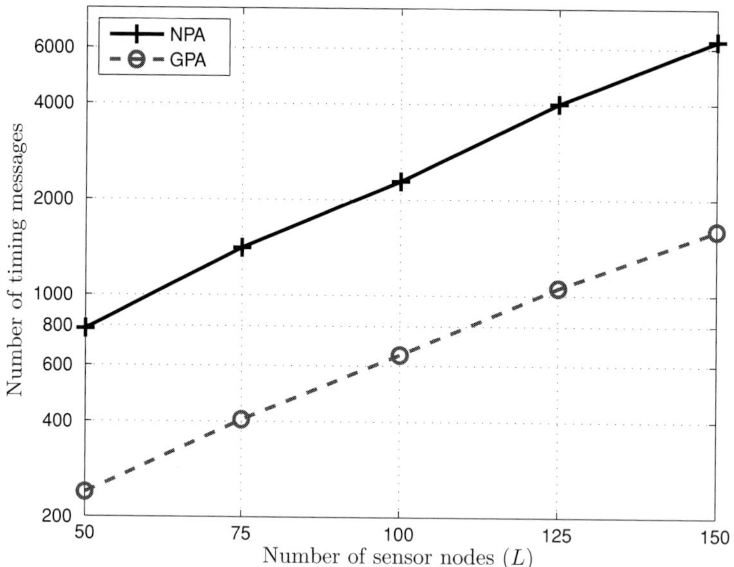

Figure 8.6: Number of messages for constructing the network hierarchy (GPA versus NPA): transmission range = 25, area = 100 × 100.

Figure 8.6 compares the complexity of NPA to establish the network connection hierarchy with that of GPA, which assumes a level hierarchy, with respect to the number of sensor nodes. In this simulation, sensors are randomly deployed in a square of size 100×100, the transmission range of each sensor is set to 25, and the reference node is assumed to be located at the center of the simulation area. It can be seen that the complexity becomes greater as the number of sensor nodes (density) increases. The number of required discovery messages for NPA is about four times larger than that for GPA. Note that here we assume a static network, i.e., the network hierarchy does not need to be reconstructed frequently. The following section analyzes the proposed algorithms in terms of the number of required timing messages, and compares them with other protocols.

8.4 COMPARISONS AND ANALYSIS

In this section we compare the proposed algorithms with conventional ones such as TPSN, RBS, and FTSP in terms of the number of required timing messages, and thus predict energy consumption for network-wide synchronization. Let $|\mathbf{p}|$ denote the number of elements in a pairwise synchronization sequence vector $\mathbf{p}$, then the total number of timing messages for NPA (N_{NPA}) is given by

$$N_{NPA} = 2N|\mathbf{p}|, \tag{8.4}$$

where N is the number of beacons per node in a pairwise synchronization. Similarly, for GPA, the total number of timing messages (N_{GPA}) is given by

$$N_{GPA} = 2N \sum_{i=1}^{N_G} |\mathbf{p}_i|, \tag{8.5}$$

where N_G denotes the number of parent–children groups and $\mathbf{p}_i$ denotes the pairwise synchronization sequence vector of the ith group. In the given example, $|\mathbf{p}| = 4$ (see Figure 8.2) and $\sum_{i=1}^{N_G} |\mathbf{p}_i| = 4$ or 6 (see Figure 8.4(a) and (b)), i.e., $N_{NPA} = 8N$ and $N_{GPA} = 8N$ or $12N$.

For multi-cluster sensor networks consisting of supernodes (the former scenario), direct comparison with the proposed PBS and RBS is not possible since they assume different network setups. For RBS, the network should be divided into a number of separate subgroups (clusters) such that every node in a subgroup is located within the transmission ranges of any other nodes in the same subgroup, i.e., a single-hop topology is applied to each subgroup. However, PBS consists of a different set of clusters, each of which is in a region covered by both supernodes. Indeed, both schemes require extra timing message exchanges between subgroups for global synchronization. There exist nodes (which could be supernodes) which are connected to multiple clusters and share global timing information to maintain network-wide synchronization. Thus, both N_{PBS} and N_{RBS} depend on the number of clusters and their connection status.

Assuming there are no supernodes in the network, the performances of N_{NPA} and N_{GPA} are compared in Figure 8.7 with those of N_{TPSN} and the lower bound for N_{RBS} with respect to the overall number of sensor nodes. Again, in this simulation, sensor nodes are randomly deployed on an area of 100×100, the transmission range of each sensor is 25, and the reference node is assumed to be located at the center of the simulation area. The number of beacons (N) is 10 in this simulation. It can be seen that PBS (with both GPA and NPA) requires a much lower number of timing messages than the other protocols, such as TPSN, FTSP, and RBS, and the gaps between the required number of message transmissions for PBS and those for the other protocols become greater as L increases. Therefore, for densely deployed WSNs, PBS has a significant benefit in terms of energy consumption over both TPSN and RBS. In addition, the performance of GPA is quite close to that of NPA even though it does not require a heuristic network connection search. As mentioned, GPA can be implemented by simply adding a group-wise connection discovery procedure to the conventional level discovery process in an arbitrary level-based synchronization protocol like TPSN.

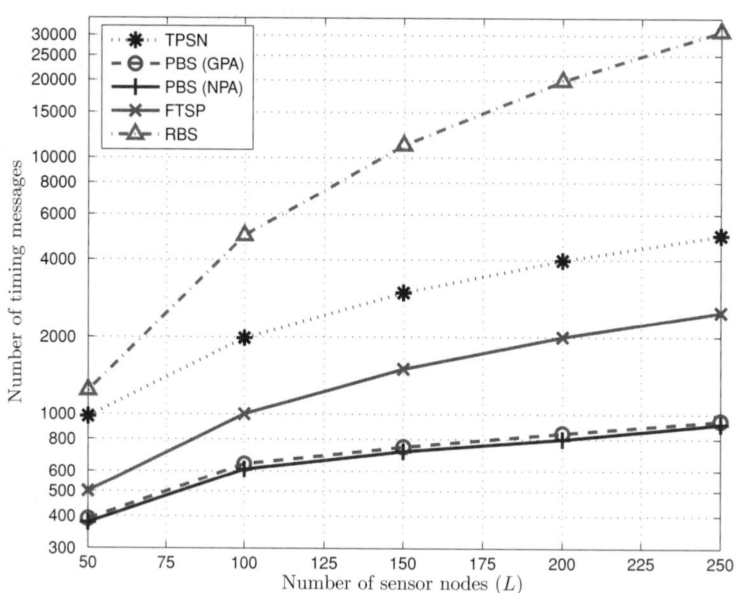

Figure 8.7: Required number of message exchanges with respect to the number of sensor nodes: transmission range = 25, area = 100 × 100, number of beacons (N) = 10.

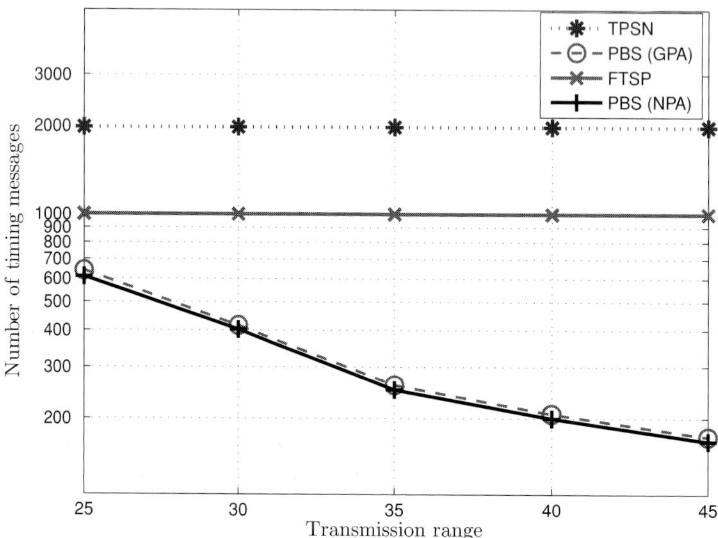

Figure 8.8: Required number of message exchanges with respect to the transmission range: number of nodes (L) = 100, area = 100 × 100, number of beacons (N) = 10.

Figure 8.8 shows the performance of the proposed algorithms with respect to the transmission range assuming the same simulation setup. The overall number of sensor nodes is fixed as 100 in this simulation. It can be seen that, as the transmission range (density of the network) increases, N_{GPA} decreases (energy efficiency increases) since a larger number of sensor nodes are able to perform ROS.

Chapter 9

ENERGY-EFFICIENT ESTIMATION OF CLOCK OFFSET FOR INACTIVE NODES

The two opposite requirements of tightly synchronizing the network with a minimum number of RF transmissions and with high accuracy can be efficiently addressed using the approach suggested by the FTSP [58], where multiple inactive nodes can hear the synchronization messages transmitted by the master node in one-way timing cells exchange mechanisms. To increase the utility of this one-way mechanism, Maroti *et al.* [58] proposed the synchronization of nodes present in the communication range of the master node (broadcasting the timing beacons), where each node receiving the timing cells transmitted by the master node estimates its own clock parameters and synchronizes with the master node accordingly. However, the similar situation pertaining to the two-way timing exchange mechanism, i.e., the framework in which the nodes located in the common broadcast region of a master and slave node can overhear the time synchronization packets between them and exploit the acquired information to achieve clock synchronization remained largely unnoticed until the PBS protocol [66], [68] introduced it (as discussed in detail in Chapter 8). Note that although the idea of SRS is quite old and has most famously been used in NTP [61] for a long time, it is due to the wireless nature of communication channels in sensornets that the technique of synchronization of silent nodes located in their common broadcast region can be exploited. Therefore, the clock synchronization requirements can be reasonably met without paying any price on the network lifetime (i.e., without exchanging additional messages for clock synchronization purposes and thereby reducing battery life) or node hardware (e.g., by improving the quality of the quartz crystals or by utilizing more expensive power-efficient batteries).

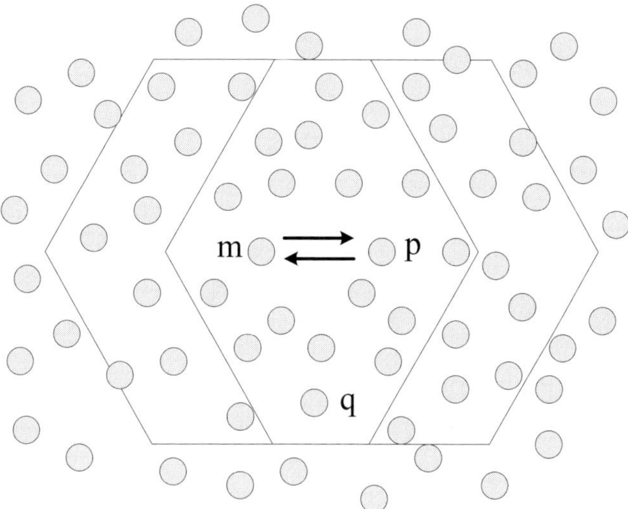

Figure 9.1: A WSN with two active nodes m and p exchanging timing cells with silent nodes like q, located within the common broadcast region of the active nodes m and p.

The main topics in this chapter are as follows. First, the MLE for the clock offset of the silent nodes, which are only receiving the timing cells exchanged by the master–slave pair is derived, and the uniqueness of the MLE is proved. One very important implication of this work is that the performance of the sender–receiver protocols, whose main disadvantage has always been categorized as the high communication overhead in WSN scenarios due to their point-to-point rather than broadcast nature can be compared with that of receiver–receiver or hybrid protocols on equal grounds. Second, the CRLBs for the clock offsets of both the active and silent nodes are derived and used as benchmarks to assess the performance of the estimators.

9.1 PROBLEM FORMULATION

Consider a WSN consisting of several sensor nodes as shown in Figure 9.1, which dynamically elect a master node m through any master election algorithm, and whose time is chosen as the *reference* time for the rest of that synchronization cycle. Depending on the SRS protocol employed for operation, node m chooses another node p as the slave node at the start of the synchronization cycle. Let ϕ_p denote the clock offset of node p with respect to node m. As illustrated in Figure 9.1, node m transmits timing cell 1 over the wireless channel to node p which responds by transmitting timing cell 2 back to node m. The timestamps $s_j^{m\rightarrow p}$ and $r_j^{p\rightarrow m}$ are

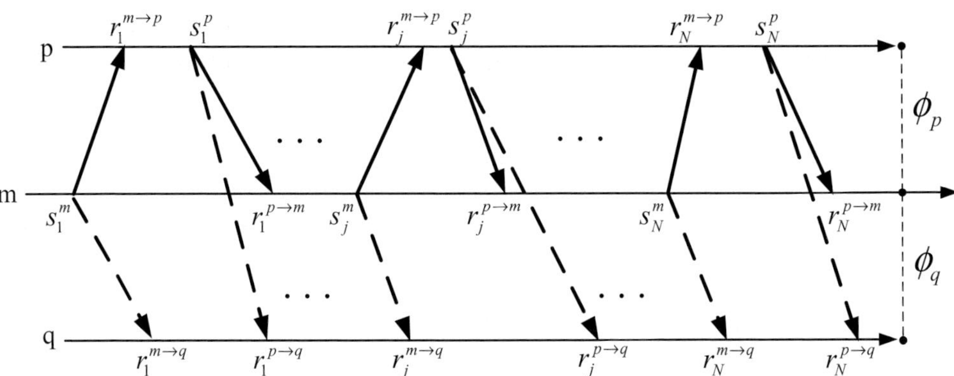

Figure 9.2: A two-way timing cell exchange mechanism between nodes m and p with node q overhearing them.

recorded by node m at pretransmission and postreception of timing cells 1 and 2, respectively. Similarly, node p records $r_j^{m\to p}$ and $s_j^{p\to m}$ according to its own time reference (offset from node m by ϕ_p) at postreception and pretransmission of timing cells 1 and 2, respectively. N such timing cells are exchanged between m and p and the first of them, $s_1^{m\to p}$, is chosen as the initial reference time.

Now observe from Figure 9.1 that if the transmission range of sensor nodes can roughly be modeled as lying within a hexagon, then a few other nodes, e.g., node q whose clock offset with respect to node m is ϕ_q, lie within the intersection of the broadcast regions of nodes m and p. Without taking part in any communication and hence conserving considerable power, node q and other similar nodes can listen to the whole message exchange *flying through the air* between nodes m and p. For this reason, let all the transmitted messages be represented by the transmitter's index *only* without any reference to the receiving node so that $s_j^{m\to p}$ and $s_j^{p\to m}$ in the above paragraph now change to s_j^m and s_j^p respectively. As illustrated in Figure 9.2, assume that node q timestamps the timing cells coming from nodes m and p as $r_j^{m\to q}$ and $r_j^{p\to q}$, respectively. Notice that node q is also receiving the packets $r_j^{m\to p}$, sent by node m and timestamped by node p, along with s_j^p because node p is required to send this information back to node m inside the packet containing s_j^p. For the discussion here, we have assumed that the deterministic part of link delays is unknown but the same for all the nodes receiving the messages from nodes m and p. This is because usually the nodes in a WSN share the same hardware specifications and characteristics and hence undergo similar transmission, reception, encoding, decoding, and byte alignment times. In addition, the propagation time of RF waveforms is less than 1 μs for ranges under 300 meters which implies that for nodes lying close by at short distances from each other, the difference in the propagation time of the same message will be even less than a few nanoseconds. Therefore, instead of $d^{m\to p}$, $d^{m\to q}$, or $d^{p\to q}$, the deterministic part of link delays is denoted as d in this chapter.

Lastly, the non-deterministic or random link delays, $z_j^{m\to p}$, $z_j^{m\to q}$ and $z_j^{p\to q}$, have been modeled as coming from an exponential distribution with similar means. A complete discussion on the justification for this modeling framework can be found in Chapter 5.

The following equations summarize the model depicted above for $j = 1, \ldots, N$.

$$\begin{aligned} r_j^{m\to p} &= s_j^m + \phi_p + d + z_j^{m\to p}, \\ r_j^{m\to q} &= s_j^m + \phi_q + d + z_j^{m\to q}, \\ r_j^{p\to q} &= s_j^p - \phi_p + \phi_q + d + z_j^{p\to q}, \end{aligned}$$

where $z_j^{m\to p}$, $z_j^{m\to q}$ and $z_j^{p\to q}$ are iid exponential RVs with the same mean λ. Rearranging the equations and introducing the notation $U_j \triangleq r_j^{m\to p} - s_j^m$, $V_j \triangleq r_j^{m\to q} - s_j^m$ and $W_j \triangleq r_j^{p\to q} - s_j^p$ yields

$$U_j = \phi_p + d + z_j^{m\to p}, \tag{9.1}$$

$$V_j = \phi_q + d + z_j^{m\to q}, \tag{9.2}$$

$$W_j = \phi_q - \phi_p + d + z_j^{p\to q}. \tag{9.3}$$

Having formulated the problem and the associated model completely, next we will present a procedure for estimating the clock offsets of these silent nodes based on the ML technique at the essentially negligible cost of a few computations.

9.2 MAXIMUM LIKELIHOOD ESTIMATION (MLE)

Based on (9.1), (9.2), and (9.3), the likelihood function can be expressed as

$$\begin{aligned} L(\lambda, d, \phi_p, \phi_q) = {} & \lambda^{-3N} \exp\left[-\frac{1}{\lambda}\sum_{j=1}^{N}\{U_j + V_j + W_j - 2\phi_q - 3d\}\right] \\ & \times \prod_{j=1}^{N} I\left[U_j - \phi_p - d\right] \prod_{j=1}^{N} I\left[V_j - \phi_q - d\right] \prod_{j=1}^{N} I\left[W_j - \phi_q + \phi_p - d\right], \end{aligned} \tag{9.4}$$

where the $3N$ unit step functions $I[\cdot]$ are defined as being equal to 1 if their argument is positive and 0 otherwise, and represent the support constraints for the likelihood function. Now since these constraints do not depend on λ, the likelihood function will be maximized by $\hat{\lambda}$ for all the fixed values of (d, ϕ_p, ϕ_q) by forcing the derivative of the log-likelihood function to be zero:

$$\frac{\partial \ln L\left(\lambda, d, \phi_p, \phi_q\right)}{\partial \lambda} = \frac{-3N}{\lambda} + \frac{1}{\lambda^2} \sum_{j=1}^{N} \left\{U_j + V_j + W_j - 2\phi_q - 3d\right\} = 0,$$

which implies

$$\hat{\lambda} = \frac{1}{3N} \sum_{j=1}^{N} \left\{U_j + V_j + W_j - 2\phi_q - 3d\right\}.$$

Plugging the above value of $\hat{\lambda}$ back into (9.4) and exploiting the fact that the indexed values in the unit step functions are independent of the unknown parameters yields the reduced likelihood function:

$$\begin{aligned} L'\left(d, \phi_p, \phi_q\right) &= \exp(-3N) \left[\frac{1}{3N} \sum_{j=1}^{N} \left\{U_j + V_j + W_j - 2\phi_q - 3d\right\}\right]^{-3N} \\ &\quad \times I\left[U_{(1)} - \phi_p - d\right] I\left[V_{(1)} - \phi_q - d\right] I\left[W_{(1)} - \phi_q + \phi_p - d\right], \end{aligned} \tag{9.5}$$

where the subscript (1) denotes the minimum order statistics of the corresponding observations, i.e., $U_{(1)}$, $V_{(1)}$, and $W_{(1)}$ are the minimum values of $\{U_j\}_{j=1}^N$, $\{V_j\}_{j=1}^N$, and $\{W_j\}_{j=1}^N$, respectively.

It is clear that the reduced likelihood function $L'(d, \phi_p, \phi_q)$ can be maximized by minimizing the expression $\sum_{j=1}^{N}\{U_j + V_j + W_j - 2\phi_q - 3d\}$, which subsequently becomes the cost function $f_0(d, \phi_q)$. Since this cost function is linear in both ϕ_q and d, the maximum cannot be found through its differentiation and hence must be searched for over the boundary of its support region. Therefore, let us closely analyze this support region by writing the constraints in the following form:

$$d > 0, \tag{9.6}$$

$$d \leq U_{(1)} - \phi_p, \tag{9.7}$$

$$d \leq V_{(1)} - \phi_q, \tag{9.8}$$

$$d \leq W_{(1)} - \phi_q + \phi_p. \tag{9.9}$$

Figure 9.3 depicts the three-dimensional support region of the reduced likelihood function over which it has to be maximized, where d is drawn as a function of ϕ_p and ϕ_q. A two-dimensional aerial view of this support region is drawn in

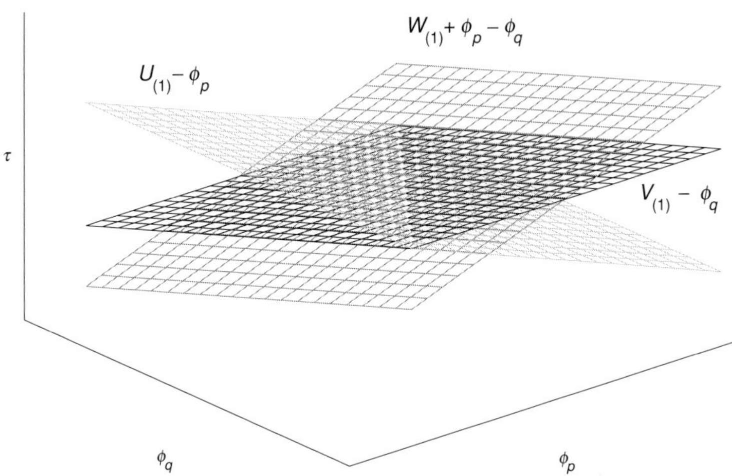

Figure 9.3: Support region of the reduced likelihood function $L'(d, \phi_p, \phi_q)$.

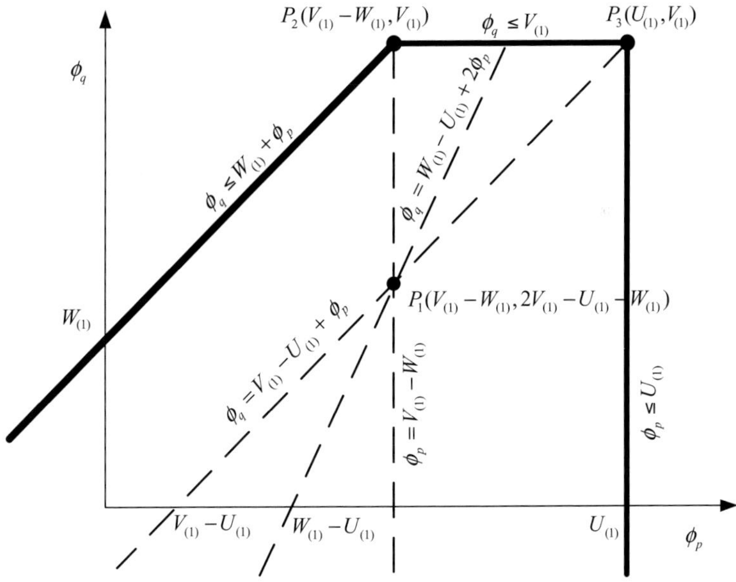

Figure 9.4: An aerial two-dimensional view of the support region.

Figure 9.4, which illustrates the lines on the (ϕ_p, ϕ_q) plane where the intersections of the curves (9.6)–(9.9) lie. Figure 9.4 is further broken down into seven regions as shown in Figure 9.5 and both Figures 9.4 and 9.5 show three sets of lines: solid, dashed, and dotted. Each of these three sets is explained in detail in the following discussion.

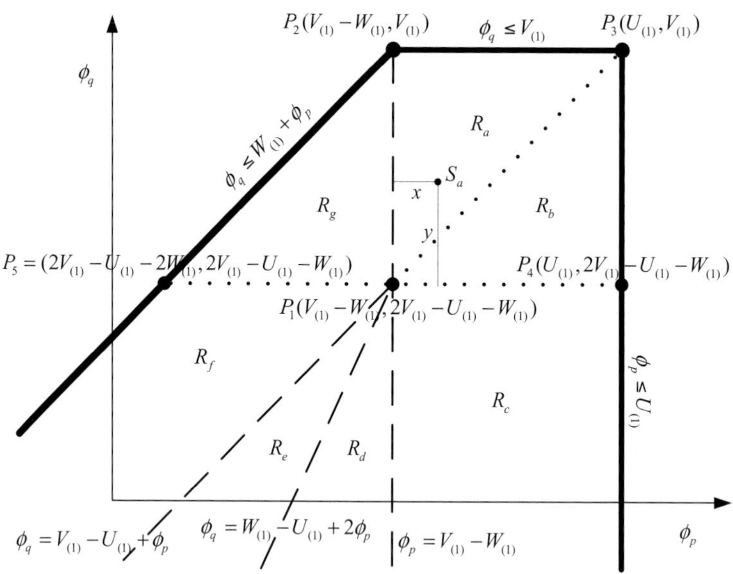

Figure 9.5: A breakdown of the support region into seven parts.

- *Solid lines* Observe that the base of this support region is formed by the intersection of (9.6) with the surfaces {(9.7), (9.8), (9.9)} respectively. Hence, slicing this three-dimensional region in Figure 9.3 horizontally at $d = 0$ reveals the two-dimensional view of this base B formed by

$$B = \begin{cases} \phi_p = U_{(1)}, & -\infty < \phi_q \leq V_{(1)} \\ \phi_q = W_{(1)} + \phi_p, & -\infty < \phi_q \leq V_{(1)}, -\infty < \phi_p \leq V_{(1)} - W_{(1)} \\ \phi_q = V_{(1)}, & V_{(1)} - W_{(1)} \leq \phi_p \leq U_{(1)} \end{cases}. \quad (9.10)$$

The border of this base B is illustrated as solid lines in Figures 9.4 and 9.5, and $f_0(d, \phi_q)$ is constrained to remain inside of it.

- *Dashed lines* As explained above and shown in Figure 9.3, the walls of the support region are formed by the three distinct surfaces (9.7), (9.8) and (9.9). The lines on the (ϕ_p, ϕ_q) plane, on which their respective intersections lie, are depicted as three dashed lines in Figures 9.4 and 9.5 and summarized in Table 9.1. Also explained by this table and shown in Figure 9.5, is the point $P_1 = (V_{(1)} - W_{(1)}, 2V_{(1)} - U_{(1)} - W_{(1)})$ on (ϕ_p, ϕ_q) plane, where all the above three surfaces meet each other, that is of paramount importance for the study considered herein.

- *Dotted lines* To simplify the derivation of the MLE and prove that it is unique, dotted lines are drawn in Figure 9.5 in order to further break the base B into easier-to-work-with geometrical figures.

Table 9.1: *Intersections of surfaces (9.7)–(9.9)*

Surfaces	Contour of intersection on (ϕ_p, ϕ_q) plane
(9.7), (9.8)	Line $\phi_q = V_{(1)} - U_{(1)} + \phi_p$
(9.8), (9.9)	Line $\phi_p = V_{(1)} - W_{(1)}$
(9.7), (9.9)	Line $\phi_q = W_{(1)} - U_{(1)} + 2\phi_p$
(9.7), (9.8), (9.9)	Point $\left(V_{(1)} - W_{(1)}, 2V_{(1)} - U_{(1)} - W_{(1)}\right)$

Note that in maximizing $L'(d, \phi_p, \phi_q)$ over the set $\Phi = \{(d, \phi_p, \phi_q) : d > 0, |\phi_p| < \infty, |\phi_q| < \infty\}$, four different cases need to be considered with reference to point P_1 and the point $P_3 = (U_{(1)}, V_{(1)})$ in Figure 9.4.

1. $\left[(V_{(1)} - W_{(1)} < U_{(1)}) \cap (2V_{(1)} - U_{(1)} - W_{(1)} < V_{(1)})\right]$: This is the case drawn in Figure 9.5 and it will suffice to derive the MLE considering it, since the other possible case is handled in a similar fashion.

2. $\left[(V_{(1)} - W_{(1)} > U_{(1)}) \cap (2V_{(1)} - U_{(1)} - W_{(1)} > V_{(1)})\right]$: In this case, boundaries of the support region and the intersections of the surfaces are drawn by a mirror image or 180° rotation of Figures 9.4 and 9.5. The MLE remains exactly the same and its derivation follows arguments similar to those in Case 1.

3. $\left[(V_{(1)} - W_{(1)} < U_{(1)}) \cap (2V_{(1)} - U_{(1)} - W_{(1)} > V_{(1)})\right]$: This case is not possible since $2V_{(1)} - U_{(1)} - W_{(1)} > V_{(1)}$ implies $V_{(1)} - W_{(1)} > U_{(1)}$, which is in contradiction with the first condition $V_{(1)} - W_{(1)} < U_{(1)}$.

4. $\left[(V_{(1)} - W_{(1)} > U_{(1)}) \cap (2V_{(1)} - U_{(1)} - W_{(1)} < V_{(1)})\right]$: This is also not possible for a similar reason to that mentioned in Case 3.

Moreover, it is clear that the MLE lies on the edge of the support region, i.e., somewhere on the ceiling of any of the surfaces (9.7)–(9.9), because if the MLE lies anywhere inside the support region at a point $C(\hat{d}, \hat{\phi}_p, \hat{\phi}_q)$, then for the same $(\hat{\phi}_p, \hat{\phi}_q)$, $f_0(\hat{d}, \hat{\phi}_q)$ can be further minimized by increasing $\hat{d}$ until it touches the edge of the overlying surface.

Having considered all the possibilities for the data and having divided the base B into the regions $R_a, R_b, \ldots, R_g$, each of these regions will be individually analyzed to derive the MLE and prove its uniqueness. From here onwards, to avoid labeling

Table 9.2: *Boundary evaluation of region R_a*

	Major properties of region R_a
Border	$V_{(1)} - W_{(1)} \leq \phi_p \cap V_{(1)} - U_{(1)} + \phi_p \leq \phi_q \leq V_{(1)}$
Coordinates of point S	$\left(V_{(1)} - W_{(1)} + x, 2V_{(1)} - U_{(1)} - W_{(1)} + y\right)$
Region characteristic	$x, y \geq 0,\ y \geq x$
Surfaces (9.7)–(9.9)	$d_U \leq U_{(1)} - V_{(1)} + W_{(1)} - x$ $d_V \leq U_{(1)} - V_{(1)} + W_{(1)} - y$ $d_W \leq U_{(1)} - V_{(1)} + W_{(1)} + x - y$
Remarks	$d_V \leq d_W$ since $x \geq 0$; $d_V \leq d_U$ since $y \geq x$
Boundary surface	$d \leq V_{(1)} - \phi_q$

too many equations and hence keeping the presentation simple, d in inequalities (9.7), (9.8), and (9.9) will be denoted by d_U, d_V and d_W, respectively.

Region R_a

Boundary evaluation: As shown in Figure 9.5, region R_a is a triangle formed by the vertices $P_1 = (V_{(1)} - W_{(1)}, 2V_{(1)} - U_{(1)} - W_{(1)})$, $P_2 = (V_{(1)} - W_{(1)}, V_{(1)})$, and $P_3 = (U_{(1)}, V_{(1)})$. To find the surface marking the boundary of this region, consider any point S_a in this region (shown in Figure 9.5) whose abscissa is at distance x from the abscissa (P_1) and ordinate is at distance y from the ordinate (P_1). Therefore, S_a is the point with coordinates $(V_{(1)} - W_{(1)} + x, 2V_{(1)} - U_{(1)} - W_{(1)} + y)$. Notice that $x \geq 0$, $y \geq 0$ is always true since both are just Euclidean distances. In addition, the relation $y \geq x$ always holds true within R_a because the point S_a lies between the lines $\phi_q = V_{(1)} - U_{(1)} + \phi_p$ and $\phi_p = V_{(1)} - W_{(1)}$. To satisfy the constraints (9.7)–(9.9) simultaneously, plug the coordinates of S_a into them such that

$$\begin{aligned} d_U &\leq U_{(1)} - V_{(1)} + W_{(1)} - x, \\ d_V &\leq U_{(1)} - V_{(1)} + W_{(1)} - y, \\ d_W &\leq U_{(1)} - V_{(1)} + W_{(1)} + x - y. \end{aligned}$$

It is clear from above that $d_V \leq d_W$ since $x \geq 0$. Also, $d_V \leq d_U$ since $y \geq x$ here. Therefore, the surface $d \leq V_{(1)} - \phi_q$ forms the boundary of the support region in R_a. The main points of the above discussion are summarized in Table 9.2.

Parameter estimation: To derive the MLE in region R_a, consider the minimization of cost function $f_0(d, \phi_q) = \sum_{j=1}^{N} \{U_j + V_j + W_j - 2\phi_q - 3d\}$. By virtue of the above

Table 9.3: *Boundary evaluation of region R_b*

	Major properties of region R_b
Border	$\phi_p \leq U_{(1)} \bigcap 2V_{(1)} - U_{(1)} - W_{(1)} \leq \phi_q \leq V_{(1)} - U_{(1)} + \phi_p$
Coordinates of point S	$\left(V_{(1)} - W_{(1)} + x, 2V_{(1)} - U_{(1)} - W_{(1)} + y\right)$
Region characteristic	$x, y \geq 0, y \leq x$
Surfaces (9.7)–(9.9)	$d_U \leq U_{(1)} - V_{(1)} + W_{(1)} - x$
	$d_V \leq U_{(1)} - V_{(1)} + W_{(1)} - y$
	$d_W \leq U_{(1)} - V_{(1)} + W_{(1)} + x - y$
Remarks	$d_V \leq d_W$ since $x \geq 0$; $d_U \leq d_V$ since $y \leq x$
Boundary surface	$d \leq U_{(1)} - \phi_p$

boundary evaluation study, the MLE lies on the surface $d \leq V_{(1)} - \phi_q$. To see the variation in $f_0(d, \phi_q)$ on this surface, substitute $d = V_{(1)} - \phi_q$ to get the modified cost function $f_0'(d, \phi_q) = \sum_{j=1}^{N}\{U_j + V_j + W_j - 3V_{(1)} + \phi_q\}$, which depends only on ϕ_q. It is clear that $f_0'(d, \phi_q)$ can be minimized by choosing $\hat{\phi}_q$ as small as possible on this particular surface, which corresponds to the point P_1 in R_a. Hence, the MLE in R_a is given by

$$\hat{\Phi}_{MLE} = \begin{bmatrix} \hat{\phi}_p \\ \hat{\phi}_q \\ \hat{d} \end{bmatrix} = \begin{bmatrix} V_{(1)} - W_{(1)} \\ 2V_{(1)} - U_{(1)} - W_{(1)} \\ U_{(1)} + W_{(1)} - V_{(1)} \end{bmatrix}. \tag{9.11}$$

Region R_b

Boundary evaluation: Following the same procedure as employed for region R_a, and summarized in Table 9.3, it is found that $d \leq U_{(1)} - \phi_p$ is the boundary surface of the support region in R_b.

Parameter estimation: As the MLE is somewhere on the edge of the support region, the MLE must lie on the surface $d \leq U_{(1)} - \phi_p$. Substituting this into the cost function $f_0(d, \phi_q)$ yields $f_0'(d, \phi_q) = \sum_{j=1}^{N}\{U_j + V_j + W_j - 3U_{(1)} + 3\phi_p - 2\phi_q\}$. Now in this case, $f_0'(d, \phi_q)$ varies on the boundary surface in R_b with both ϕ_p and ϕ_q, where the minimum ϕ_p (due to the positive sign) corresponds to the point P_1, but the maximum ϕ_q (due to the negative sign) corresponds to the point P_3.

To derive the MLE, consider a point S_b anywhere in the region R_b at a distance of $\sqrt{x^2 + y^2}$ from the point P_1 and with the coordinates $(V_{(1)} - W_{(1)} + x,$

$2V_{(1)} - U_{(1)} - W_{(1)} + y)$. It is evident that within this region, $x \geq y$. Now relating $f_0'(d, \phi_q)$ to the point S_b through the boundary surface yields

$$
\begin{aligned}
f_0'(d, \phi_q) &= \sum_{j=1}^{N} \Big\{ U_j + V_j + W_j - 3U_{(1)} + 3(V_{(1)} - W_{(1)} + x) \\
&\qquad - 2(2V_{(1)} - U_{(1)} - W_{(1)} + y) \Big\} \\
&\propto 3x - 2y.
\end{aligned}
$$

The maximum value y can achieve in R_b is x, which implies $3x - 2y \geq 3x - 2x = x$. Hence, the minimization problem of $f_0'(d, \phi_q)$ is equivalent to minimization of $3x - 2y$, which in turn is proportional to minimization of x. It is clear from Figure 9.5 that x achieves its minimum value at point P_1. It can also be verified by considering the region R_b as a sum of vertical segments starting on the line $\phi_q = 2V_{(1)} - U_{(1)} - W_{(1)}$ and ending on the line $\phi_q = V_{(1)} - U_{(1)} + \phi_p$, with infinitesimal distances between them. Since ϕ_p is constant on each such vertical line segment, $f_0'(d, \phi_q)$ can be minimized by the greatest possible ϕ_q, which coincides with the line $\phi_q = V_{(1)} - U_{(1)} + \phi_p$. This gives a set of points on this line for which the minimum should be searched, which in turn can be found by noting that $f_0'(d, \phi_q)$ is proportional to $3\phi_p - 2\phi_p = \phi_p$ on the line $\phi_q = V_{(1)} - U_{(1)} + \phi_p$, which corresponds to the minimum ϕ_p, and hence the point P_1.

Therefore, the MLE in R_b is the same as in R_a and is given by the expression in (9.11).

Region R_c

Boundary evaluation: Working along similar lines as before, Table 9.4 summarizes the boundary evaluation problem in region R_c. The boundary surface of the support region here is enveloped by $d \leq U_{(1)} - \phi_p$.

Parameter estimation: Finding the MLE in region R_c is straightforward. From the above boundary evaluation study, $f_0'(d, \phi_q)$ is given by $\sum_{j=1}^{N}\{U_j + V_j + W_j - 3U_{(1)} + 3\phi_p - 2\phi_q\}$. Clearly, this can be minimized by making ϕ_p as small as possible and ϕ_q as large as possible, both of which conditions are satisfied by the point P_1. Hence, the MLE in R_b is again given by (9.11).

Region R_d

Boundary evaluation: As summarized in Table 9.5, the boundary surface in region R_d is $d \leq U_{(1)} - \phi_p$.

Table 9.4: *Boundary evaluation of region* R_c

	Major properties of region R_c
Border	$V_{(1)} - W_{(1)} \leq \phi_p \leq U_{(1)} \cap \phi_q \leq 2V_{(1)} - U_{(1)} - W_{(1)}$
Coordinates of point S	$\left(V_{(1)} - W_{(1)} + x, 2V_{(1)} - U_{(1)} - W_{(1)} - y\right)$
Region characteristic	$x, y \geq 0,$
Surfaces (9.7)–(9.9)	$d_U \leq U_{(1)} - V_{(1)} + W_{(1)} - x$
	$d_V \leq U_{(1)} - V_{(1)} + W_{(1)} + y$
	$d_W \leq U_{(1)} - V_{(1)} + W_{(1)} + x + y$
Remarks	$d_V \leq d_W$ since $x \geq 0$; $d_U \leq d_V$ since $x, y \geq 0$
Boundary surface	$d \leq U_{(1)} - \phi_p$

Table 9.5: *Boundary evaluation of region* R_d

	Major properties of region R_d
Border	$\phi_p \leq V_{(1)} - W_{(1)} \cap \phi_q \leq W_{(1)} - U_{(1)} + 2\phi_p$
Coordinates of point S	$\left(V_{(1)} - W_{(1)} - x, 2V_{(1)} - U_{(1)} - W_{(1)} - y\right)$
Region characteristic	$x, y \geq 0,\ y \geq 2x$
Surfaces (9.7)–(9.9)	$d_U \leq U_{(1)} - V_{(1)} + W_{(1)} + x$
	$d_V \leq U_{(1)} - V_{(1)} + W_{(1)} + y$
	$d_W \leq U_{(1)} - V_{(1)} + W_{(1)} - x + y$
Remarks	$d_W \leq d_V$ since $x \geq 0$; $d_U \leq d_W$
	since $y \geq x + x \Rightarrow y - x \geq x$
Boundary surface	$d \leq U_{(1)} - \phi_p$

Parameter estimation: In region R_d, again $f_0'(d, \phi_q)$ is proportional to $3\phi_p - 2\phi_q$. Although the maximum ϕ_q corresponds to the point P_1, the minimum ϕ_p does not, requiring a closer look at the region. Now consider a point S_d anywhere in R_d whose abscissa and ordinate are $V_{(1)} - W_{(1)} - x$ and $2V_{(1)} - U_{(1)} - W_{(1)} - y$, respectively. Over the point S_d and its neighborhood, $f_0'(d, \phi_q)$ is given by

$$\begin{aligned} f_0'(d, \phi_q) &= \sum_{j=1}^{N} \Big\{ U_j + V_j + W_j - 3U_{(1)} + 3(V_{(1)} - W_{(1)} - x) \\ &\qquad - 2(2V_{(1)} - U_{(1)} - W_{(1)} - y) \Big\}, \\ &\propto -3x + 2y. \end{aligned}$$

Table 9.6: *Boundary evaluation of region R_e*

	Major properties of region R_e
Border	$W_{(1)} - U_{(1)} + 2\phi_p \leq \phi_q \leq V_{(1)} - U_{(1)} + \phi_p$
Coordinates of point S	$\left(V_{(1)} - W_{(1)} - x, 2V_{(1)} - U_{(1)} - W_{(1)} - y\right)$
Region characteristic	$x, y \geq 0,\ x \leq y \leq 2x$
Surfaces (9.7)–(9.9)	$d_U \leq U_{(1)} - V_{(1)} + W_{(1)} + x$
	$d_V \leq U_{(1)} - V_{(1)} + W_{(1)} + y$
	$d_W \leq U_{(1)} - V_{(1)} + W_{(1)} - x + y$
Remarks	$d_U \leq d_V$ since $x \leq y$; $d_W \leq d_U$
	since $y \leq x + x \Rightarrow y - x \leq x$
Boundary surface	$d \leq W_{(1)} + \phi_p - \phi_q$

Table 9.7: *Boundary evaluation of region R_f*

	Major properties of region R_f
Border	$V_{(1)} - U_{(1)} + \phi_p \leq \phi_q \leq W_{(1)} \cdots$
	$\cdots + \phi_p \bigcap \phi_q \leq 2V_{(1)} - U_{(1)} - W_{(1)}$
Coordinates of point S	$\left(V_{(1)} - W_{(1)} - x, 2V_{(1)} - U_{(1)} - W_{(1)} - y\right)$
Region characteristic	$x, y \geq 0,\ y \leq x$
Surfaces (9.7)–(9.9)	$d_U \leq U_{(1)} - V_{(1)} + W_{(1)} + x$
	$d_V \leq U_{(1)} - V_{(1)} + W_{(1)} + y$
	$d_W \leq U_{(1)} - V_{(1)} + W_{(1)} - x + y$
Remarks	$d_V \leq d_U$ since $y \leq x$; $d_W \leq d_V$ since $x \geq 0$
Boundary surface	$d \leq W_{(1)} + \phi_p - \phi_q$

Since $y \geq 2x$ in R_d and the minimum value y can achieve is $2x$, $-3x + 2y \geq -3x + 2(2x) = x$. Therefore, minimization of $f_0'(d, \phi_q)$ corresponds to minimization of $-3x + 2y$ which requires minimization of x. Recall that $x, y \geq 0$, resulting in the coordinates of point P_1 being the MLE for (ϕ_p, ϕ_q).

Regions R_e and R_f

Boundary evaluation: Tables 9.6 and 9.7 show that the surface $d \leq W_{(1)} + \phi_p - \phi_q$ is the envelope of the support region in both R_e and R_f. Notice that these two regions could have been combined as one larger region because both the boundary surface and the MLE (as shown in the next subsection) are the same for R_f and R_e. This has not been pursued due to the difference in the boundary evaluation procedure, since $d_W \le d_U \le d_V$ in R_e, but $d_W \le d_V \le d_U$ in R_f.

Parameter estimation: In these two regions, the MLE lies on the surface $d \le W_{(1)} + \phi_p - \phi_q$, which is plugged into $f_0(d, \phi_q)$ to yield $f_0'(d, \phi_q) = \sum_{j=1}^N \{U_j + V_j + W_j - 3W_{(1)} - 3\phi_p + \phi_q\}$. Again, the maximum ϕ_p (owing to the negative sign) in these two regions yields the point P_1 as the solution. However, the minimum ϕ_q (owing to the positive sign) corresponds to the open areas of R_e and R_f where $\phi_q \to -\infty$. Therefore, consider a point S_{ef} somewhere in either of these two regions with the coordinates $(V_{(1)} - W_{(1)} - x, 2V_{(1)} - U_{(1)} - W_{(1)} - y)$. Over this point S_{ef} and in its vicinity, $f_0'(d, \phi_q)$ can be written as

$$
\begin{aligned}
f_0'(d, \phi_q) &= \sum_{j=1}^{N} \Big\{ U_j + V_j + W_j - 3W_{(1)} - 3(V_{(1)} - W_{(1)} - x) \\
&\qquad + (2V_{(1)} - U_{(1)} - W_{(1)} - y) \Big\} \\
&\propto 3x - y.
\end{aligned}
$$

Note that minimizing $f_0'(d, \phi_q)$ is now equivalent to minimizing the expression $3x - y$. Using the relationship $y \le 2x$ in these two regions, x can achieve a minimum value of $y/2$ which implies $3x - y \ge 3y/2 - y = y/2$. A positive coefficient, $1/2$, with y above implies that it should be chosen as small as possible, which is achieved on point P_1. Therefore, the MLE in these two cases is also given by (9.11).

Region R_g

Boundary evaluation: From Table 9.8, it is clear that the boundary surface on R_g is $d \le W_{(1)} + \phi_p - \phi_q$.

Parameter estimation: In this region, the modified cost function $f_0'(d, \phi_q)$ is again proportional to the expression $-3\phi_p + \phi_q$. It is evident that ϕ_p should be maximized and ϕ_q should be minimized for the minimization of $f_0'(d, \phi_q)$, both of which can be accomplished by choosing the point P_1.

In conclusion, the MLE $(\hat{\phi}_p, \hat{\phi}_q, \hat{d})$ for each of regions $R_a - R_g$ is given by (9.11), and hence it is unique.

In the next section, we turn our attention to deriving the CRLB for any unbiased estimator of the clock offsets ϕ_p and ϕ_q.

Table 9.8: *Boundary evaluation of region R_g*

	Major properties of region R_g
Border	$\phi_p \leq V_{(1)} - W_{(1)} \bigcap 2V_{(1)} - U_{(1)} \cdots$ $\cdots - W_{(1)} \leq \phi_q \leq W_{(1)} + \phi_p$
Coordinates of point S	$\left(V_{(1)} - W_{(1)} - x, 2V_{(1)} - U_{(1)} - W_{(1)} + y\right)$
Region characteristic Surfaces (9.7)–(9.9)	$x, y \geq 0,$ $d_U \leq U_{(1)} - V_{(1)} + W_{(1)} + x$ $d_V \leq U_{(1)} - V_{(1)} + W_{(1)} - y$ $d_W \leq U_{(1)} - V_{(1)} + W_{(1)} - x - y$
Remarks	$d_W \leq d_U$ since $x, y \geq 0$; $d_W \leq d_V$ since $x \geq 0$
Boundary surface	$d \leq W_{(1)} + \phi_p - \phi_q$

9.3 CRAMER–RAO LOWER BOUND (CRLB)

In practical applications, it is extremely useful to know in advance the best performance an estimator might achieve by deriving a lower bound for it. In addition to providing information on how well the estimator can perform, it helps the researchers to find an unbiased estimator that has the minimum possible variance among all unbiased estimators. Also, it gives a benchmark against which different estimators can be compared to rank the finest one(s), without going through an empirical procedure. In this particular problem, finding the CRLB is helpful for both $\hat{\phi}_p$ and $\hat{\phi}_q$. For $\hat{\phi}_q$, it can obviously set the performance benchmark for any unbiased estimator of clock offset when a node like q is silently listening to the timing cell exchange between a pair of nodes in the vicinity; whereas for $\hat{\phi}_p$, it can compare whether the clock offset of an active node like p, estimated by an inactive node like q, can perform better than the one which node p itself can estimate during a two-way timing cell exchange with reference node m using the observations s_j^m, $r_j^{m \rightarrow p}$, s_j^p and $r_j^{p \rightarrow m}$ (derived in [41]). If that is indeed the case, then any of the inactive nodes, say q, can transmit this new estimate $\hat{\phi}_p$ to node p for improved performance, albeit at the cost of one extra communication.

The CRLB theorem states that if the likelihood function satisfies certain regularity conditions [44], the variance of any unbiased estimator $\hat{\theta}$ must satisfy the relationship

$$\mathrm{var}(\hat{\theta}) \geq I^{-1}(\theta),$$

where $I(\theta)$ is the quantity known as Fisher information and is defined as

$$I(\theta) = -E\left[\frac{\partial^2 \ln L(\theta)}{\partial \theta^2}\right] = E\left[\left(\frac{\partial \ln L(\theta)}{\partial \theta}\right)^2\right].$$

Clearly, the domain of the likelihood function (a product of independent pdfs and hence a pdf itself) depends on both unknown parameters ϕ_p and ϕ_q due to which the order of differentiation and integration in the regularity condition cannot be interchanged and hence CRLB cannot be found by employing the likelihood function. However, there is an alternative technique available for deriving the CRLB which exploits the pdf of the estimator itself as explained below.

9.3.1 CRLB FOR THE CLOCK OFFSET OF INACTIVE NODE $\hat{\phi}_q$

Working on $\hat{\phi}_q$ first, note that from (9.11)

$$\begin{aligned} \hat{\phi}_q &= 2V_{(1)} - U_{(1)} - W_{(1)} \\ &= 2\left(d + \phi_q + z_{(1)}^{m\to q}\right) - \left(d + \phi_p + z_{(1)}^{m\to p}\right) - \left(d + \phi_q - \phi_p + z_{(1)}^{p\to q}\right) \\ &= \phi_q + 2z_{(1)}^{m\to q} - z_{(1)}^{m\to p} - z_{(1)}^{p\to q}. \end{aligned} \tag{9.12}$$

Notice that $\hat{\phi}_q$ is an unbiased function of ϕ_q, since

$$E\left[\hat{\phi}_q\right] = E\left[2V_{(1)} - U_{(1)} - W_{(1)}\right] = \phi_q + 2\frac{\lambda}{N} - \frac{\lambda}{N} - \frac{\lambda}{N} = \phi_q,$$

and its variance is

$$\operatorname{var}\left(\hat{\phi}_q\right) = E\left[\left(\hat{\phi}_q - \phi_q\right)^2\right] = E\left[\left(2z_{(1)}^{m\to q} - z_{(1)}^{m\to p} - z_{(1)}^{p\to q}\right)^2\right] = 6\frac{\lambda^2}{N^2}, \tag{9.13}$$

where the fact that the first-order statistics $z_{(1)}^{m\to p}$, $z_{(1)}^{m\to q}$, and $z_{(1)}^{p\to q}$ are also exponential RVs with mean λ/N and variance λ^2/N^2 has been used. The pdf of $\hat{\phi}_q$ can be derived as follows. Consider (9.12) which can be written as

$$\hat{\phi}_q - \phi_q = 2z_{(1)}^{m\to q} - \left(z_{(1)}^{m\to p} + z_{(1)}^{p\to q}\right) = g - h, \tag{9.14}$$

where $g = 2z_{(1)}^{m\to q}$ and $h = z_{(1)}^{m\to p} + z_{(1)}^{p\to q}$ for simplicity. It is straightforward to show that the pdf of the first-order statistic $z_{(1)}^{m\to p}$ from the observation set $\{z_j^{m\to p}\}_{j=1}^N$ (and correspondingly that of $z_{(1)}^{m\to q}$ and $z_{(1)}^{p\to q}$) is given as

$$\begin{aligned} f_{z_{(1)}^{m\to p}}\left(z_{(1)}^{m\to p}\right) &= N\left[1 - F_{z_j^{m\to p}}\left(z_{(1)}^{m\to p}\right)\right]^{N-1} f_{z_j^{m\to p}}\left(z_{(1)}^{m\to p}\right) \\ &= \frac{N}{\lambda}\exp\left[-\frac{N}{\lambda}z_{(1)}^{m\to p}\right] \qquad z_{(1)}^{m\to p} \geq 0, \end{aligned} \tag{9.15}$$

where $f_{z_j^{m\to p}}$ and $F_{z_j^{m\to p}}$ are the pdf and cumulative distribution function (CDF) of the exponential random variables $z_j^{m\to p}$, respectively. Therefore, the pdfs of the first-order statistics $z_{(1)}^{m\to p}$, $z_{(1)}^{m\to q}$, and $z_{(1)}^{p\to q}$ are also exponential with mean λ/N. Now turning to (9.14) and using (9.15), it is clear that

$$f_G(g) = \frac{N}{2\lambda}\exp\left[-\frac{N}{2\lambda}g\right] I\left[g\right]. \tag{9.16}$$

Since $z_{(1)}^{m\to p}$ and $z_{(1)}^{p\to q}$ are the first-order statistics of independent data sets $\{z_j^{m\to p}\}_{j=1}^N$ and $\{z_j^{p\to q}\}_{j=1}^N$, respectively, these are also independent with the distribution (9.15). To find the pdf of h, note that

$$\begin{aligned} f_H(h) &= \int_{-\infty}^{\infty} f_{z_{(1)}^{m\to p}}\left(h - z_{(1)}^{p\to q}\right) f_{z_{(1)}^{p\to q}}\left(z_{(1)}^{p\to q}\right) I\left[h - z_{(1)}^{p\to q}\right] I\left[z_{(1)}^{p\to q}\right] dz_{(1)}^{p\to q} \\ &= \frac{N^2}{\lambda^2}\int_0^h \exp\left[-\frac{N}{\lambda}\left(h - z_{(1)}^{p\to q}\right)\right]\exp\left[-\frac{N}{\lambda}z_{(1)}^{p\to q}\right] dz_{(1)}^{p\to q} \\ &= \frac{N^2}{\lambda^2}h\exp\left[-\frac{N}{\lambda}h\right] I\left[h\right], \end{aligned} \tag{9.17}$$

which is a Gamma distribution with shape parameter 2 and scale parameter λ/N. We conclude that $\hat{\phi}_q - \phi_q$ is equal to the difference between an exponential RV and a Gamma, both of which are independent and positive valued. Therefore, $g - h$ can acquire any value from $-\infty$ to ∞ and the final pdf of $\hat{\phi}_q$ can be derived using (9.16) and (9.17) as follows.

For $\hat{\phi}_q \le \phi_q$, we have

$$\begin{aligned} f_{\hat{\phi}_q}\left(\hat{\phi}_q\right) &= \frac{N^3}{2\lambda^3}\int_{-\infty}^{\infty}\exp\left[-\frac{N}{2\lambda}\left(\hat{\phi}_q - \phi_q + h\right)\right] I\left[\hat{\phi}_q - \phi_q + h\right] h\exp\left[-\frac{N}{\lambda}h\right] I\left[h\right] dh \\ &= \frac{N^3}{2\lambda^3}\exp\left[-\frac{N}{2\lambda}\left(\hat{\phi}_q - \phi_q\right)\right]\int_{-(\hat{\phi}_q - \phi_q)}^{\infty} h\exp\left[-\frac{3N}{2\lambda}h\right] dh \\ &= \frac{N^3}{2\lambda^3}\exp\left[-\frac{N}{2\lambda}\left(\hat{\phi}_q - \phi_q\right)\right]\left[\exp\left[\frac{3N}{2\lambda}\left(\hat{\phi}_q - \phi_q\right)\right]\left(-\frac{2\lambda(\hat{\phi}_q - \phi_q)}{3N} + \frac{4\lambda^2}{9N^2}\right)\right] \\ &= \frac{2N}{9\lambda}\exp\left[\frac{N}{\lambda}\left(\hat{\phi}_q - \phi_q\right)\right] - \frac{N^2}{3\lambda^2}\left(\hat{\phi}_q - \phi_q\right)\exp\left[\frac{N}{\lambda}\left(\hat{\phi}_q - \phi_q\right)\right] \\ &= \frac{N}{3\lambda}\left[\frac{2}{3} - \frac{N}{\lambda}\left(\hat{\phi}_q - \phi_q\right)\right]\exp\left[\frac{N}{\lambda}\left(\hat{\phi}_q - \phi_q\right)\right]. \end{aligned}$$

And for $\hat{\phi}_q \geq \phi_q$, we infer that

$$\begin{aligned} f_{\hat{\phi}_q}\left(\hat{\phi}_q\right) &= \frac{N^3}{2\lambda^3}\int_{-\infty}^{\infty} \exp\left[-\frac{N}{2\lambda}\left(\hat{\phi}_q - \phi_q + h\right)\right] h \exp\left[-\frac{N}{\lambda}h\right] I\left[h\right] I\left[\hat{\phi}_q - \phi_q + h\right] dh \\ &= \frac{N^3}{2\lambda^3}\exp\left[-\frac{N}{2\lambda}\left(\hat{\phi}_q - \phi_q\right)\right]\int_0^{\infty} h\exp\left[-\frac{3N}{2\lambda}h\right] dh \\ &= \frac{N^3}{2\lambda^3}\exp\left[-\frac{N}{2\lambda}\left(\hat{\phi}_q - \phi_q\right)\right]\left[\frac{4\lambda^2}{9N^2}\right] \\ &= \frac{2N}{9\lambda}\exp\left[-\frac{N}{2\lambda}\left(\hat{\phi}_q - \phi_q\right)\right]. \end{aligned}$$

Therefore, the pdf of $f_{\hat{\phi}_q}\left(\hat{\phi}_q\right)$ can now be expressed as

$$f_{\hat{\phi}_q}\left(\hat{\phi}_q\right) = \begin{cases} \frac{N}{3\lambda}\left[\frac{2}{3} - \frac{N}{\lambda}\left(\hat{\phi}_q - \phi_q\right)\right]\exp\left[\frac{N}{\lambda}\left(\hat{\phi}_q - \phi_q\right)\right] & \hat{\phi}_q \leq \phi_q \\ \frac{2N}{9\lambda}\exp\left[-\frac{N}{2\lambda}\left(\hat{\phi}_q - \phi_q\right)\right] & \hat{\phi}_q \geq \phi_q \end{cases}.$$

To check if it is indeed a valid pdf, note that

$$\begin{aligned} \frac{2N}{9\lambda}\int_{-\infty}^{0}\exp\left[\frac{N}{\lambda}\left(\hat{\phi}_q - \phi_q\right)\right] d\left(\hat{\phi}_q - \phi_q\right) &= \frac{2}{9}, \\ -\frac{N^2}{3\lambda^2}\int_{-\infty}^{0}\left(\hat{\phi}_q - \phi_q\right)\exp\left[\frac{N}{\lambda}\left(\hat{\phi}_q - \phi_q\right)\right] d\left(\hat{\phi}_q - \phi_q\right) &= \frac{1}{3}, \\ \frac{2N}{9\lambda}\int_{0}^{\infty}\exp\left[-\frac{N}{2\lambda}\left(\hat{\phi}_q - \phi_q\right)\right] d\left(\hat{\phi}_q - \phi_q\right) &= \frac{4}{9}, \end{aligned}$$

which sum up to 1. Finally, to verify its unbiasedness, note that

$$\begin{aligned} E\left[\hat{\phi}_q\right] &= \frac{N}{3\lambda}\int_{-\infty}^{\phi_q}\hat{\phi}_q\left[\frac{2}{3} - \frac{N}{\lambda}\left(\hat{\phi}_q - \phi_q\right)\right]\exp\left[\frac{N}{\lambda}\left(\hat{\phi}_q - \phi_q\right)\right] d\hat{\phi}_q \\ &\quad + \frac{2N}{9\lambda}\int_{\phi_q}^{\infty}\hat{\phi}_q\exp\left[-\frac{N}{2\lambda}\left(\hat{\phi}_q - \phi_q\right)\right] d\hat{\phi}_q \\ &= \left(-\frac{N}{3\lambda}{\phi_q}^2 + \frac{2}{3}\phi_q - \frac{2\lambda}{3N} + \frac{2}{9}\phi_q + \frac{N}{3\lambda}{\phi_q}^2 - \frac{1}{3}\phi_q - \frac{2\lambda}{9N}\right) + \left(\frac{4}{9}\phi_q + \frac{8\lambda}{9N}\right) \\ &= \phi_q. \end{aligned}$$

Clearly, this is not differentiable at the point $\hat{\phi}_q = \phi_q$, but exploiting its continuity at this point ($f_{\hat{\phi}_q}\left(\phi_q+\right) = f_{\hat{\phi}_q}\left(\phi_q-\right) = 2N/9\lambda$), its domain is independent of ϕ_q. Differentiating $\ln f_{\hat{\phi}_q}\left(\hat{\phi}_q\right)$ with respect to ϕ_q yields

$$\frac{\partial \ln f_{\hat{\phi}_q}\left(\hat{\phi}_q\right)}{\partial \phi_q} = \begin{cases} \frac{N}{\lambda\left[\frac{2}{3} - \frac{N}{\lambda}\left(\hat{\phi}_q - \phi_q\right)\right]} - \frac{N}{\lambda} & \hat{\phi}_q \leq \phi_q \\ \frac{N}{2\lambda} & \hat{\phi}_q \geq \phi_q \end{cases}. \tag{9.18}$$

Taking its expected value results in

$$
\begin{aligned}
E\left[\frac{\partial \ln f_{\hat{\phi}_q}(\hat{\phi}_q)}{\partial \phi_q}\right] &= \int_{-\infty}^{\phi_q} \frac{N}{\lambda}\frac{N}{3\lambda} \exp\left[\frac{N}{\lambda}(\hat{\phi}_q - \phi_q)\right] d\hat{\phi}_q \\
&\quad - \int_{-\infty}^{\phi_q} \frac{N}{\lambda}\frac{N}{3\lambda}\left[\frac{2}{3} - \frac{N}{\lambda}(\hat{\phi}_q - \phi_q)\right] \exp\left[\frac{N}{\lambda}(\hat{\phi}_q - \phi_q)\right] d\hat{\phi}_q \\
&\quad + \int_{\phi_q}^{\infty} \frac{N}{2\lambda}\frac{2N}{9\lambda} \exp\left[-\frac{N}{2\lambda}(\hat{\phi}_q - \phi_q)\right] d\hat{\phi}_q \\
&= \frac{N}{3\lambda}\int_{-\infty}^{\phi_q} \left[\frac{N}{3\lambda} + \frac{N^2(\hat{\phi}_q - \phi_q)}{\lambda^2}\right] \exp\left[\frac{N}{\lambda}(\hat{\phi}_q - \phi_q)\right] d\hat{\phi}_q \\
&\quad + \frac{N^2}{9\lambda^2}\int_{\phi_q}^{\infty} \exp\left[-\frac{N}{2\lambda}(\hat{\phi}_q - \phi_q)\right] d\hat{\phi}_q \\
&= -\frac{2N}{9\lambda} + \frac{2N}{\lambda} = 0.
\end{aligned}
$$

Having satisfied both the requirements (unbiasedness and the regularity condition), we can now proceed to derive the CRLB. Differentiating (9.18) again with respect to ϕ_q,

$$
\frac{\partial^2 \ln f_{\hat{\phi}_q}(\hat{\phi}_q)}{\partial \phi_q^2} = \begin{cases} -\frac{N^2}{\lambda^2}\left[\frac{2}{3} - \frac{N}{\lambda}(\hat{\phi}_q - \phi_q)\right]^{-2} & \hat{\phi}_q \le \phi_q \\ 0 & \hat{\phi}_q \ge \phi_q \end{cases} .
$$

Taking the expectation on both sides gives

$$
E\left[\frac{\partial^2 \ln f_{\hat{\phi}_q}(\hat{\phi}_q)}{\partial \phi_q^2}\right] = -\frac{N^3}{3\lambda^3}\int_{-\infty}^{\phi_q}\left[\frac{2}{3} - \frac{N}{\lambda}(\hat{\phi}_q - \phi_q)\right]^{-1} \exp\left[\frac{N}{\lambda}(\hat{\phi}_q - \phi_q)\right] d\hat{\phi}_q .
$$

A change of variable $t = N/\lambda\,(\hat{\phi}_q - \phi_q) - 2/3$ leads to

$$
\begin{aligned}
E\left[\frac{\partial^2 \ln f_{\hat{\phi}_q}(\hat{\phi}_q)}{\partial \phi_q^2}\right] &= \frac{N^2}{3\lambda^2} \exp\left(\frac{2}{3}\right) \int_{-\infty}^{-2/3} t^{-1}\exp(t)dt \\
&= \frac{N^2}{3\lambda^2} \exp\left(\frac{2}{3}\right) \mathrm{Ei}(-2/3) \qquad (9.19) \\
&= -0.258664\,\frac{N^2}{\lambda^2},
\end{aligned}
$$

where $\mathrm{Ei}(x)$ is the well-known *exponential integral function*, which is defined as

$$
\mathrm{Ei}(x) = \begin{cases} -\int_{-x}^{\infty} t^{-1}\exp(-t)\,dt = \int_{-\infty}^{x} t^{-1}\exp(t)\,dt, & x < 0 \\ -\lim\limits_{\epsilon\to+0}\left[\int_{-x}^{-\epsilon} t^{-1}\exp(-t)\,dt + \int_{\epsilon}^{\infty} t^{-1}\exp(-t)\,dt\right], & x > 0 \end{cases} .
$$

In (9.19) above, the value of $\mathrm{Ei}(-2/3)$ has been computed as $-0.398\,409$ and $\exp\left(\frac{2}{3}\right) = 1.947\,734$. Therefore, the CRLB for ϕ_q is given by the expression

$$\mathrm{CRLB}\left(\hat{\phi}_q\right) = 3.866\frac{\lambda^2}{N^2}. \tag{9.20}$$

Note that the variance of $\hat{\phi}_q$ is inversely proportional to the square of the number of observations N^2 and hence decreases very rapidly as the nodes exchange more messages. Having derived the Fisher information, observe from (9.18) that

$$\begin{aligned}\frac{\partial \ln f_{\hat{\phi}_q}\left(\hat{\phi}_q\right)}{\partial \phi_q} &\neq I\left(\phi_q\right)\left(\hat{\phi}_q - \phi_q\right), \\ &= \frac{N^2}{3.866\lambda^2}\left(2V_{(1)} - U_{(1)} - W_{(1)} - \phi_q\right).\end{aligned}$$

Two points are worth commenting on here. First, the MLE is not efficient, since it does not satisfy the relation an efficient estimator necessarily fulfills. Second, an efficient estimator for the problem targeted in this chapter does *not* exist owing to the rule: *if an efficient estimator exists, the maximum likelihood procedure will produce it*. Consequently, it is shown in Chapter 10 that for symmetric delays no unbiased estimator can be found having a lower variance than the MLE, and hence it is also the MVUE.

9.3.2 CRLB FOR THE CLOCK OFFSET OF ACTIVE NODE $\hat{\phi}_p$

Since the domain of the likelihood function in (9.4) depends on ϕ_p, it cannot be utilized for finding the CRLB. Working with the pdf of $\hat{\phi}_p$ using (9.11), and proceeding in a similar way as before, we have

$$\begin{aligned}\hat{\phi}_p &= V_{(1)} - W_{(1)} = d + \phi_q + z_{(1)}^{m\to q} - \left(d + \phi_q - \phi_p + z_{(1)}^{p\to q}\right), \\ &= \phi_p + z_{(1)}^{m\to q} - z_{(1)}^{p\to q}.\end{aligned}$$

The mean and variance of $\hat{\phi}_p$, are given respectively by

$$\begin{aligned}E\left[\hat{\phi}_p\right] &= E\left[\phi_p + z_{(1)}^{m\to q} - z_{(1)}^{p\to q}\right] = \phi_p + \frac{\lambda}{N} - \frac{\lambda}{N} = \phi_p, \\ E\left[\left(\hat{\phi}_p - \phi_p\right)^2\right] &= E\left[\left(z_{(1)}^{m\to q} - z_{(1)}^{p\to q}\right)^2\right] = 2\frac{\lambda^2}{N^2}.\end{aligned}$$

Having confirmed the unbiasedness of this estimator, since the difference between two exponential RVs with mean λ/N is a Laplacian random variable with mean 0, $f_{\hat{\phi}_p}(\phi_p)$ can be written as

$$f_{\hat{\phi}_p}(\phi_p) = \begin{cases} \frac{N}{2\lambda} \exp\left[\frac{N}{\lambda}(\hat{\phi}_p - \phi_p)\right] & \hat{\phi}_p \leq \phi_p \\ \frac{N}{2\lambda} \exp\left[-\frac{N}{\lambda}(\hat{\phi}_p - \phi_p)\right] & \hat{\phi}_p \geq \phi_p \end{cases}.$$

It is evident that the pdf of $\hat{\phi}_p$ is symmetric around ϕ_p which implies that $E[(\partial \ln f_{\hat{\phi}_p}(\phi_p)/\partial\phi_p)] = 0$. Differentiating both sides with respect to ϕ_p and taking the expectation of its square,

$$E\left[\left(\frac{\partial \ln f_{\hat{\phi}_p}(\phi_p)}{\partial \phi_p}\right)^2\right] = \frac{N^2}{\lambda^2},$$

and hence the CRLB for $\hat{\phi}_p$ can be expressed as

$$\text{CRLB}(\hat{\phi}_p) = \frac{\lambda^2}{N^2},$$

where again the variance is inversely proportional to N^2. Since this CRLB is slightly greater than the CRLB for $\hat{\phi}_p$ derived as $\lambda^2/4N^2$ in (6.23) (where $\alpha = \lambda$), it can be concluded that instead of the silent nodes like q estimating ϕ_p and communicating this estimate to node p, node p should estimate $\hat{\phi}_p$ by itself using the two-way timing message exchange with the reference node m.

9.4 Simulation Results

Computer simulations have been performed to illustrate the MSE (or variance, since the estimators are unbiased) and CRLB for the estimators $\hat{\phi}_p$ and $\hat{\phi}_q$, where the mean of the exponential link delays has been chosen as 1. Figure 9.6 shows this comparison on a logarithmic scale where the MSE of both estimators decreases with the square of the number of observations. This is due to the positive-only nature of the link delays justifiably modeled as exponential RVs. Had these delays been obeying a symmetric pdf, such as a Gaussian, the MSE would have decreased proportionally with the number of data points, instead of a decrease proportional to the square of the number of data points. In addition, notice from Figure 9.6 that there is a constant difference between the MSE of the active and inactive nodes due to the plot being drawn on a logarithmic scale. If the curves are plotted on a normal scale instead,

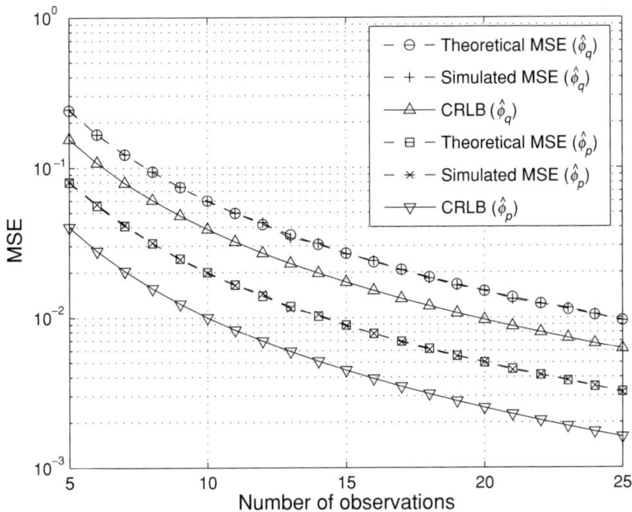

Figure 9.6: Simulations illustrating the MSE of $\hat{\phi}_q$ and $\hat{\phi}_p$ on a logarithmic scale for $\lambda = 1$.

the difference between the MSE diminishes as the number of observations increases since the clock offset estimators of both types of nodes are inversely proportional to N^2. In conclusion, the scheme is an attractive choice, even more so when its cost-free quality is taken into account.

Chapter 10

SOME IMPROVED AND GENERALIZED ESTIMATION SCHEMES FOR CLOCK SYNCHRONIZATION OF INACTIVE NODES

In Chapter 9, the MLE for the clock offset and mean link delays of the inactive node were derived using the symmetric exponential delay model. This chapter not only presents better estimation techniques than the MLE, but also addresses the problem using the more realistic asymmetric delay model.

The main topics in this chapter are as follows. First, the generalized least-squares theory is applied on the order statistics of the synchronization packets to obtain the BLUE-OS of the clock offsets of the inactive node and the mean link delays, which is the optimal solution in the class of linear unbiased estimators. Second, the restriction of the estimates to being linear is then removed through the derivation of the MVUE of the clock offset. Third, since the MSE may be decreased by adding a little bias to the estimator with the potential effect of more significantly reducing the estimator's variance relative to the increased bias, the MMSE estimator with the expected loss independent of clock offset and fixed delay is also obtained, thus further improving the synchronization quality.

Based on the same model with equal α, β, and γ, the MLE derived in the Chapter 9 is expressed as

$$\hat{\mathbf{\Phi}}_S = \begin{bmatrix} \hat{\phi}_q \\ \hat{\phi}_p \\ \hat{d} \end{bmatrix} = \begin{bmatrix} 2V_{(1)} - U_{(1)} - W_{(1)} \\ V_{(1)} - W_{(1)} \\ U_{(1)} - V_{(1)} + W_{(1)} \end{bmatrix}, \tag{10.1}$$

where the subscript S points to the estimates being driven for symmetric link delays and the subscript (1) denotes the minimum order statistics of their respective data sets.

10.1 Asymmetric Exponential Link Delays

In most communications and wireless channels, and ad-hoc networks with time-varying topologies, the network delays are asymmetric in nature. Therefore, a study for deriving the efficient estimators in this case is of paramount importance. Let the order statistics of the observations $\{U_k\}_{k=1}^N$, $\{V_k\}_{k=1}^N$, and $\{W_k\}_{k=1}^N$ be denoted as $\{U_{(k)}\}_{k=1}^N$, $\{V_{(k)}\}_{k=1}^N$, and $\{W_{(k)}\}_{k=1}^N$, respectively. Transforming the data set as

$$
\begin{aligned}
U'_k &\triangleq \frac{1}{\alpha}\left(U_k - \phi_p - d\right), \\
V'_k &\triangleq \frac{1}{\beta}\left(V_k - \phi_q - d\right), \\
W'_k &\triangleq \frac{1}{\gamma}\left(W_k - \phi_q + \phi_p - d\right)
\end{aligned}
$$

makes it a set of independent observations on the standardized variate and hence the distribution becomes parameter-free. The order statistics of U'_k, V'_k, and W'_k are denoted by $U'_{(k)}$, $V'_{(k)}$, and $W'_{(k)}$, respectively. Now it is straightforward to see that

$$
\begin{aligned}
E\left[U_{(k)}\right] = \phi_p + d + \alpha E\left[U'_{(k)}\right], \text{var}\left[U_{(k)}\right] &= \alpha^2 \text{var}\left[\mathrm{U}'_{(\mathrm{k})}\right], \\
\text{cov}\left[U_{(k)}U_{(j)}\right] &= \alpha^2 \text{cov}\left[U'_{(k)}U'_{(j)}\right], \\
E\left[V_{(k)}\right] = \phi_q + d + \beta E\left[V'_{(k)}\right], \text{var}\left[V_{(k)}\right] &= \beta^2 \text{var}\left[\mathrm{V}'_{(\mathrm{k})}\right], \\
\text{cov}\left[V_{(k)}V_{(j)}\right] &= \beta^2 \text{cov}\left[V'_{(k)}V'_{(j)}\right], \\
E\left[W_{(k)}\right] = \phi_q - \phi_p + d + \gamma E\left[W'_{(k)}\right], \text{var}\left[W_{(k)}\right] &= \gamma^2 \text{var}\left[\mathrm{W}'_{(\mathrm{k})}\right], \\
\text{cov}\left[W_{(k)}W_{(j)}\right] &= \gamma^2 \text{cov}\left[W'_{(k)}W'_{(j)}\right].
\end{aligned}
$$

Next, the statistics of the ordered samples (see [42]) can be expressed as

$$
\begin{aligned}
E\left[U'_{(k)}\right] &= E\left[V'_{(k)}\right] = E\left[W'_{(k)}\right] = \sum_{i=1}^{k} \frac{1}{(N-i+1)}, \\
\text{var}\left[U'_{(k)}\right] &= \text{var}\left[V'_{(k)}\right] = \text{var}\left[W'_{(k)}\right] = \sum_{i=1}^{k} \frac{1}{(N-i+1)^2}, \\
\text{cov}\left[U'_{(k)}U'_{(j)}\right] &= \text{cov}\left[V'_{(k)}V'_{(j)}\right] = \text{cov}\left[W'_{(k)}W'_{(j)}\right] = \sum_{i=1}^{k} \frac{1}{(N-i+1)^2}.
\end{aligned}
$$

Therefore, the $N \times N$ symmetric positive-definite covariance matrix $\mathbf{C}$ for the vectors $[U_{(1)}, \ldots, U_{(N)}]$, $[V_{(1)}, \ldots, V_{(N)}]$, and $[W_{(1)}, \ldots, W_{(N)}]$ takes the form

$$\mathbf{C} = \begin{bmatrix} \frac{1}{N^2} & \frac{1}{N^2} & \cdots & \frac{1}{N^2} \\ \frac{1}{N^2} & \frac{1}{N^2} + \frac{1}{(N-1)^2} & \cdots & \frac{1}{N^2} + \frac{1}{(N-1)^2} \\ \vdots & \vdots & \cdots & \vdots \\ \frac{1}{N^2} & \frac{1}{N^2} + \frac{1}{(N-1)^2} & \cdots & \sum_{k=1}^{N} \frac{1}{(N-k+1)^2} \end{bmatrix}.$$

As explained before, the inverse of this covariance matrix can be found by the application of Gauss–Jordan elimination as

$$\mathbf{C}^{-1} = \begin{bmatrix} N^2 + (N-1)^2 & -(N-1)^2 & 0 & \cdots & 0 \\ -(N-1)^2 & (N-1)^2 + (N-2)^2 & -(N-2)^2 & \cdots & 0 \\ 0 & -(N-2)^2 & (N-2)^2 + (N-3)^2 & \cdots & 0 \\ \vdots & \vdots & \vdots & \cdots & \vdots \\ 0 & 0 & 0 & \cdots & 1 \end{bmatrix}.$$

Now we proceed towards estimating the clock parameters and mean link delays as follows.

10.1.1 BEST LINEAR UNBIASED ESTIMATION USING ORDER STATISTICS (BLUE-OS)

It is well known that the derivation of regular BLUE in general yields suboptimal results, since the class of unbiased estimators, within which the search is performed, is restricted to be linear. In the case in which the noise is normally distributed, the direct application of BLUE provides the optimal solution by virtue of the Gauss–Markov theorem [44]. Therefore, for distributions other than Gaussian, including the exponential distribution as is the case with the modeling framework adopted in this chapter, the application of BLUE is not of much significance given its lack of optimality. However, for a general location-scale distribution, Lloyd [55] suggested a new technique based on the derivation of BLUE using order statistics instead of just the raw observations. This technique is used in the scenario addressed in this chapter.

Let $\mathbf{\Phi}_A \triangleq [\phi_q \ \phi_p \ d \ \alpha \ \beta \ \gamma]^T$, where the subscript A denotes the relevance of estimators to asymmetric link delays and $\mathbf{z} \triangleq [U_{(1)} \ U_{(2)} \cdots U_{(N)} \ V_{(1)} \ V_{(2)} \cdots V_{(N)} \ W_{(1)} \ W_{(2)} \cdots W_{(N)}]^T$. Then the linear model based on the ordered observations can be expressed as

$$E[\mathbf{z}] = \begin{bmatrix} \mathbf{C}_1 & \mathbf{C}_2 & \mathbf{C}_3 \end{bmatrix}^T \mathbf{\Phi}_A = \mathbf{Q}\mathbf{\Phi}_A,$$

where the matrices $\mathbf{C}_1$, $\mathbf{C}_2$, and $\mathbf{C}_3$ are given by

$$\mathbf{C}_1 = \begin{bmatrix} 0 & 0 & \cdots & 0 \\ 1 & 1 & \cdots & 1 \\ 1 & 1 & \cdots & 1 \\ \frac{1}{N} & \frac{1}{N}+\frac{1}{N-1} & \cdots & \sum_{k=1}^{N}\frac{1}{(N-k+1)} \\ 0 & 0 & \cdots & 0 \\ 0 & 0 & \cdots & 0 \end{bmatrix}, \quad \mathbf{C}_2 = \begin{bmatrix} 1 & 1 & \cdots & 1 \\ 0 & 0 & \cdots & 0 \\ 1 & 1 & \cdots & 1 \\ 0 & 0 & \cdots & 0 \\ \frac{1}{N} & \frac{1}{N}+\frac{1}{N-1} & \cdots & \sum_{k=1}^{N}\frac{1}{(N-k+1)} \\ 0 & 0 & \cdots & 0 \end{bmatrix},$$

$$\mathbf{C}_3 = \begin{bmatrix} 1 & 1 & \cdots & 1 \\ -1 & -1 & \cdots & -1 \\ 1 & 1 & \cdots & 1 \\ 0 & 0 & \cdots & 0 \\ 0 & 0 & \cdots & 0 \\ \frac{1}{N} & \frac{1}{N}+\frac{1}{N-1} & \cdots & \sum_{k=1}^{N}\frac{1}{(N-k+1)} \end{bmatrix}.$$

Notice that $\mathbf{Q}$ is a known matrix of dimensions $3N \times 6$, and $\mathbf{\Phi}_A$ is the 6×1 vector of unknown parameters. Since the model has been shown to be linear in terms of the ordered observations, the BLUE can be expressed as

$$\hat{\mathbf{\Phi}}_A = \left(\mathbf{Q}^T\mathbf{C}_\mathbf{z}^{-1}\mathbf{Q}\right)^{-1}\mathbf{Q}^T\mathbf{C}_\mathbf{z}^{-1}\mathbf{z}, \tag{10.2}$$

where $\mathbf{C}_\mathbf{z}$ is the covariance matrix for vector $\mathbf{z}$. Due to the mutual independence of $U_{(k)}$, $V_{(k)}$, and $W_{(k)}$, $\mathbf{C}_\mathbf{z}$ is a diagonal matrix:

$$\mathbf{C}_\mathbf{z} = \begin{bmatrix} \alpha^2\mathbf{C} & \mathbf{0} & \mathbf{0} \\ \mathbf{0} & \beta^2\mathbf{C} & \mathbf{0} \\ \mathbf{0} & \mathbf{0} & \gamma^2\mathbf{C} \end{bmatrix},$$

and its inverse can be expressed as

$$\mathbf{C}_\mathbf{z}^{-1} = \begin{bmatrix} \alpha^{-2}\mathbf{C}^{-1} & \mathbf{0} & \mathbf{0} \\ \mathbf{0} & \beta^{-2}\mathbf{C}^{-1} & \mathbf{0} \\ \mathbf{0} & \mathbf{0} & \gamma^{-2}\mathbf{C}^{-1} \end{bmatrix}.$$

Based on the above expression, it follows that

$$\mathbf{Q}^T\mathbf{C}_\mathbf{z}^{-1}\mathbf{Q} =$$
$$\begin{bmatrix} \left(\beta^{-2}+\gamma^{-2}\right)N^2 & -\gamma^{-2}N^2 & \left(\beta^{-2}+\gamma^{-2}\right)N^2 & 0 & \beta^{-2}N & \gamma^{-2}N \\ -\gamma^{-2}N & \left(\alpha^{-2}+\gamma^{-2}\right)N^2 & \left(\alpha^{-2}-\gamma^{-2}\right)N^2 & \alpha^{-2}N & 0 & -\gamma^{-2}N \\ \left(\beta^{-2}+\gamma^{-2}\right)N^2 & \left(\alpha^{-2}-\gamma^{-2}\right)N^2 & \left(\alpha^{-2}+\beta^{-2}+\gamma^{-2}\right)N^2 & \alpha^{-2}N & \beta^{-2}N & \gamma^{-2}N \\ 0 & \alpha^{-2}N & \alpha^{-2}N & \alpha^{-2}N & 0 & 0 \\ \beta^{-2}N & 0 & \beta^{-2}N & 0 & \beta^{-2}N & 0 \\ \gamma^{-2}N & -\gamma^{-2}N & \gamma^{-2}N & 0 & 0 & \gamma^{-2}N \end{bmatrix}$$

and its inverse takes the form

$$\left(\mathbf{Q}^T\mathbf{C}_\mathbf{z}^{-1}\mathbf{Q}\right)^{-1} = \begin{bmatrix} \mathbf{E}_1 & \mathbf{E}_2 \\ \mathbf{E}_3 & \mathbf{E}_4 \end{bmatrix}, \tag{10.3}$$

where

$$\mathbf{E}_1 = \frac{1}{N(N-1)} \begin{bmatrix} \alpha^2+4\beta^2+\gamma^2 & 2\beta^2+\gamma^2 & -\left(\alpha^2+2\beta^2+\gamma^2\right) \\ 2\beta^2+\gamma^2 & \beta^2+\gamma^2 & -\left(\beta^2+\gamma^2\right) \\ -\left(\alpha^2+2\beta^2+\gamma^2\right) & -\left(\beta^2+\gamma^2\right) & \alpha^2+\beta^2+\gamma^2 \end{bmatrix},$$

$$\mathbf{E}_2 = \frac{1}{N(N-1)} \begin{bmatrix} \alpha^2 & -2\beta^2 & \gamma^2 \\ 0 & -\beta^2 & \gamma^2 \\ -\alpha^2 & \beta^2 & -\gamma^2 \end{bmatrix}, \quad \mathbf{E}_3 = \frac{1}{N(N-1)} \begin{bmatrix} \alpha^2 & 0 & -\alpha^2 \\ -2\beta^2 & -\beta^2 & \beta^2 \\ \gamma^2 & \gamma^2 & -\gamma^2 \end{bmatrix},$$

$$\mathbf{E}_4 = \frac{1}{(N-1)} \begin{bmatrix} \alpha^2 & 0 & 0 \\ 0 & \beta^2 & 0 \\ 0 & 0 & \gamma^2 \end{bmatrix}.$$

Consequently,

$$\left(\mathbf{Q}^T\mathbf{C}_\mathbf{z}^{-1}\mathbf{Q}\right)^{-1}\mathbf{Q}^T\mathbf{C}_\mathbf{z}^{-1} = \frac{1}{N(N-1)} \begin{bmatrix} \mathbf{D}_1 & \mathbf{D}_2 & \mathbf{D}_3 \end{bmatrix}, \tag{10.4}$$

where the matrices $\mathbf{D}_1$, $\mathbf{D}_2$, and $\mathbf{D}_3$ are defined as follows

$$\mathbf{D}_1 = \begin{bmatrix} -\left(N^2-1\right) & 1 & \cdots & 1 \\ 0 & 0 & \cdots & 0 \\ N^2-1 & -1 & \cdots & -1 \\ -N(N-1) & N & \cdots & N \\ 0 & 0 & \cdots & 0 \\ 0 & 0 & \cdots & 0 \end{bmatrix}, \quad \mathbf{D}_2 = \begin{bmatrix} 2\left(N^2-1\right) & -2 & \cdots & -2 \\ N^2-1 & -1 & \cdots & -1 \\ -\left(N^2-1\right) & 1 & \cdots & 1 \\ 0 & 0 & \cdots & 0 \\ -N(N-1) & N & \cdots & N \\ 0 & 0 & \cdots & 0 \end{bmatrix},$$

$$\mathbf{D}_3 = \begin{bmatrix} -\left(N^2-1\right) & 1 & \cdots & 1 \\ -\left(N^2-1\right) & 1 & \cdots & 1 \\ N^2-1 & -1 & \cdots & -1 \\ 0 & 0 & \cdots & 0 \\ 0 & 0 & \cdots & 0 \\ -N(N-1) & N & \cdots & N \end{bmatrix}.$$

Plugging (10.4) into (10.2), straightforward computations lead to the following closed form for the BLUE-OS:

$$\hat{\mathbf{\Phi}}_A = \frac{1}{N-1}\begin{bmatrix} N\left(2V_{(1)} - U_{(1)} - W_{(1)}\right) - (2\overline{V} - \overline{U} - \overline{W}) \\ N\left(V_{(1)} - W_{(1)}\right) - (\overline{V} - \overline{W}) \\ N\left(U_{(1)} - V_{(1)} + W_{(1)}\right) - (\overline{U} - \overline{V} + \overline{W}) \\ N\left(\overline{U} - U_{(1)}\right) \\ N\left(\overline{V} - V_{(1)}\right) \\ N\left(\overline{W} - W_{(1)}\right) \end{bmatrix}. \quad (10.5)$$

10.1.2 Minimum Variance Unbiased Estimation (MVUE)

The ultimate goal in parameter estimation is often to find the estimator that achieves the minimum MSE, which explains why MMSE is usually adopted as the criterion of performance in most practical applications. However, it is well known in theory that the optimal MSE estimators are not realizable in general. The MSE for an arbitrary parameter θ assumes the following expression:

$$\begin{aligned} \text{MSE}(\hat{\theta}) &= E\left[(\hat{\theta} - \theta)^2\right] \\ &= \text{var}(\hat{\theta}) + \text{bias}^2(\hat{\theta}). \end{aligned}$$

It is evident that the MSE is composed of two components, namely the estimator variance and squared bias. In light of the above considerations, a technique chosen to attain realizable yet best estimators is to constrain the bias to be zero (since the dependence of the MMSE estimator on the unknown parameter typically comes from the bias). Therefore, restricting the class of estimators to be unbiased and then finding the estimator with the smallest variance for all values of the unknown parameter yields the optimal solution within the class of unbiased estimators. Hence, we proceed towards deriving the MVUE for the clock offset and mean link delays for the problem at hand.

The MVUE in the current scenario is being obtained by following the steps resulting from the Rao–Blackwell–Lehmann–Scheffé theorem, as explained in detail in Chapter 5. In the asymmetric delays case, the likelihood function for the clock offset as a function of observations $\{U_k\}_{k=1}^N$, $\{V_k\}_{k=1}^N$, and $\{W_k\}_{k=1}^N$ from (9.1), (9.2), and (9.3) is given by

$$
\begin{aligned}
L\left(\phi_q, \phi_p, d, \alpha, \beta, \gamma\right) &= (\alpha\beta\gamma)^{-N} \\
&\times \exp\left\{-\sum_{k=1}^{N}\left[\frac{1}{\alpha}\left(U_k - \phi_p - d\right) + \frac{1}{\beta}\left(V_k - \phi_q - d\right) + \frac{1}{\gamma}\left(W_k - \phi_q + \phi_p - d\right)\right]\right\} \\
&\times I\left[U_{(1)} - \phi_p - d\right] I\left[V_{(1)} - \phi_q - d\right] I\left[W_{(1)} - \phi_q + \phi_p - d\right],
\end{aligned}
\tag{10.6}
$$

where $I[\cdot]$ denotes the unit step function. Exploiting the fact that the raw sample mean and the ordered sample mean are actually the same, (10.6) can be factored as

$$
\begin{aligned}
L\left(\phi_q, \phi_p, d, \alpha, \beta, \gamma\right) &= g_1\left(\sum_{k=1}^{N} U_{(k)}, \phi_p, \alpha\right) g_2\left(\sum_{k=1}^{N} V_{(k)}, \phi_q, \beta\right) g_3\left(\sum_{k=1}^{N} W_{(k)}, \phi_q, \phi_p, \gamma\right) \\
&\times g_4\left(U_{(1)}, \phi_p, d\right) g_5\left(V_{(1)}, \phi_q, d\right) g_6\left(W_{(1)}, \phi_q, \phi_p, d\right) h_1(d, \alpha, \beta, \gamma),
\end{aligned}
$$

where

$$
\begin{aligned}
g_1\left(\sum_{k=1}^{N} U_{(k)}, \phi_p, \alpha\right) &= \exp\left[-\frac{1}{\alpha}\sum_{k=1}^{N}\left(U_{(k)} - \phi_p\right)\right], \\
g_2\left(\sum_{k=1}^{N} V_{(k)}, \phi_q, \beta\right) &= \exp\left[-\frac{1}{\beta}\sum_{k=1}^{N}\left(V_{(k)} - \phi_q\right)\right], \\
g_3\left(\sum_{k=1}^{N} W_{(k)}, \phi_q, \phi_p, \gamma\right) &= \exp\left[-\frac{1}{\gamma}\sum_{k=1}^{N}\left(W_{(k)} - \phi_q + \phi_p\right)\right], \\
g_4\left(U_{(1)}, \phi_p, d\right) &= I\left[U_{(1)} - \phi_p - d\right], \\
g_5\left(V_{(1)}, \phi_q, d\right) &= I\left[V_{(1)} - \phi_q - d\right], \\
g_6\left(W_{(1)}, \phi_q, \phi_p, d\right) &= I\left[W_{(1)} - \phi_q + \phi_p - d\right], \\
h_1\left(d, \alpha, \beta, \gamma\right) &= (\alpha\beta\gamma)^{-N} \exp\left(Nd\left[\frac{1}{\alpha} + \frac{1}{\beta} + \frac{1}{\gamma}\right]\right).
\end{aligned}
$$

In the above relations, $g_1(\sum_{k=1}^{N} U_{(k)}, \phi_p, \alpha)$, $g_2(\sum_{k=1}^{N} V_{(k)}, \phi_q, \beta)$, $g_3(\sum_{k=1}^{N} W_{(k)}, \phi_q, \phi_p, \gamma)$, $g_4(U_{(1)}, \phi_p, d)$, $g_5(V_{(1)}, \phi_q, d)$, and $g_6(W_{(1)}, \phi_q, \phi_p, d)$ are functions depending on the data only through

$$
\mathbf{T} = \left(\sum_{k=1}^{N} U_{(k)}, U_{(1)}, \sum_{k=1}^{N} V_{(k)}, V_{(1)}, \sum_{k=1}^{N} W_{(k)}, W_{(1)}\right).
$$

Therefore, according to the Neymann–Fisher factorization theorem, $\mathbf{T}$ is a sufficient statistic for $\mathbf{\Phi}_A$.

Since $\dim(\mathbf{T}) = \dim(\mathbf{\Phi}_A)$, we have to find a 6×1 vector function $\hat{\mathbf{\Phi}}_A$ such that $E[\hat{\mathbf{\Phi}}_A] = \mathbf{\Phi}_A$, provided that $\mathbf{T}$ is a complete sufficient statistic. Since the pdf of $\mathbf{T}$ is required to check whether $\mathbf{T}$ is complete, and $\sum_{k=1}^{N} U_{(k)}$ and $U_{(1)}$, $\sum_{k=1}^{N} V_{(k)}$ and $V_{(1)}$, and $\sum_{k=1}^{N} W_{(k)}$ and $W_{(1)}$ are not independent, we proceed as follows.

Taking into account only the data set $\{V_{(k)}\}_{k=1}^{N}$, it is evident that the pdf of the minimum order statistic $V_{(1)}$ is exponential with mean β/N, whereas the joint pdf of $V_{(1)}, V_{(2)}, \ldots, V_{(N)}$ is given by

$$f\left(V_{(1)}, V_{(2)}, \ldots, V_{(N)}\right) = N!\beta^{-N} \exp\left(-\frac{1}{\beta}\sum_{k=1}^{N}\left\{V_{(k)} - \phi_q - d\right\}\right)\prod_{k=1}^{N} I\left[V_{(k)} - \phi_q - d\right]. \tag{10.7}$$

Now consider the transformation ([42]),

$$\eta_k = (N-k+1)\left(V_{(k)} - V_{(k-1)}\right), \quad k = 1, 2, \ldots, N,$$

where $V_{(0)} = \phi_q + d$. Since $\sum_{k=1}^{N}(V_{(k)} - \phi_q - d) = \sum_{k=1}^{N}\eta_k$ and the Jacobian of the transformation is $N!$, substitution in (10.7) reveals that

$$p(\eta_1, \eta_2, \ldots, \eta_N) = \beta^{-N}\exp\left(-\frac{1}{\beta}\sum_{k=1}^{N}\eta_k\right)\prod_{k=1}^{N} I[\eta_k],$$

i.e., η_k are *independent* exponential random variables with similar mean β. In addition, since each $\eta_k \sim \exp(\beta)$, each η_k also assumes a Gamma distribution $\eta_k \sim \Gamma(1, \beta)$. Using the relationship $\sum_{k=1}^{N}(V_{(k)} - V_{(1)}) = \sum_{k=2}^{N}\eta_k$, and the fact that each of $\eta_2, \eta_3, \ldots, \eta_N$ is independent of η_1 (and hence of $V_{(1)}$, since $\eta_1 = N(V_{(1)} - \phi_q - d)$), $\sum_{k=1}^{N}(V_{(k)} - V_{(1)}) \sim \Gamma(N-1, \beta)$ and is independent of $V_{(1)}$.

By a similar reasoning, it can be inferred that $\sum_{k=1}^{N}(U_{(k)} - U_{(1)}) \sim \Gamma(N-1, \alpha)$ and $\sum_{k=1}^{N}(W_{(k)} - W_{(1)}) \sim \Gamma(N-1, \gamma)$ and are independent of $U_{(1)}$ and $W_{(1)}$, respectively. Therefore, the one-to-one function $\mathbf{T}' = \{\sum_{k=1}^{N}(U_{(k)} - U_{(1)}), U_{(1)}, \sum_{k=1}^{N}(V_{(k)} - V_{(1)}), V_{(1)}, \sum_{k=1}^{N}(W_{(k)} - W_{(1)}), W_{(1)}\}$ of $\mathbf{T}$ is also sufficient for estimating $\mathbf{\Phi}_A$ because the sufficient statistics are unique within one-to-one transformations. Consequently, $\mathbf{T}'$ consists of six independent random variables, which in terms of the three-parameter Gamma distribution are given by

$$u = \sum_{k=1}^{N}(U_{(k)} - U_{(1)}) \sim \Gamma(N-1, \alpha, 0), \quad U_{(1)} \sim \Gamma(1, \alpha/N, \phi_p + d),$$

$$v = \sum_{k=1}^{N}(V_{(k)} - V_{(1)}) \sim \Gamma(N-1, \beta, 0), \quad V_{(1)} \sim \Gamma(1, \beta/N, \phi_q + d),$$

$$w = \sum_{k=1}^{N}(W_{(k)} - W_{(1)}) \sim \Gamma(N-1, \gamma, 0), \quad W_{(1)} \sim \Gamma(1, \gamma/N, \phi_q - \phi_p + d).$$

Note that the domains of u, v and w are controlled by $U_{(1)}$, $V_{(1)}$, and $W_{(1)}$, respectively. Next, it has to be checked whether $\mathbf{T}'$, or equivalently $\mathbf{T}$, is complete. Completeness implies that there is only one function of $\mathbf{T}$ that is unbiased. Let $g(\mathbf{T}')$ be a function of $\mathbf{T}'$ such that $E[g(\mathbf{T}')] = \mathbf{\Phi}_A$. Suppose that there exists another function h for which $E[h(\mathbf{T}')] = \mathbf{\Phi}_A$ is also true. Then,

$$E\left[g\left(\mathbf{T}'\right) - h\left(\mathbf{T}'\right)\right] = E\left[\pi\left(\mathbf{T}'\right)\right] = 0, \qquad \forall\ \mathbf{\Phi}_A$$

where $\pi(\mathbf{T}') \triangleq g(\mathbf{T}') - h(\mathbf{T}')$ and the expectation is taken with respect to $p(\mathbf{T}'; \mathbf{\Phi}_A)$. As a result,

$$\begin{aligned}
&\int\!\!\int\!\!\int\!\!\int\!\!\int\!\!\int_{R_{\{U_{(1)},V_{(1)},W_{(1)}\}}} \pi\left(u, U_{(1)}, v, V_{(1)}, w, W_{(1)}\right) \frac{\alpha^{-(N-1)}}{\Gamma(N-1)} u^{N-2} \exp\left(-\frac{u}{\alpha}\right) \frac{N}{\alpha} \\
&\times \exp\left(-\frac{N}{\alpha}\left\{U_{(1)} - \phi_p - d\right\}\right) \frac{\beta^{-(N-1)}}{\Gamma(N-1)} v^{N-2} \exp\left(-\frac{v}{\beta}\right) \frac{N}{\beta} \\
&\times \exp\left(-\frac{N}{\beta}\left\{V_{(1)} - \phi_q - d\right\}\right) \frac{\gamma^{-(N-1)}}{\Gamma(N-1)} w^{N-2} \exp\left(-\frac{w}{\gamma}\right) \frac{N}{\gamma} \\
&\times \exp\left(-\frac{N}{\gamma}\left\{W_{(1)} - \phi_q + \phi_p - d\right\}\right) \\
&\times du\, dU_{(1)}\, dv\, dV_{(1)}\, dw\, dW_{(1)} \;=\; 0, \qquad \forall\ \mathbf{\Phi}_A,
\end{aligned}$$

where $R_{U_{(1)},V_{(1)},W_{(1)}}$ is the region defined by $I[U_{(1)} - \phi_p - d]$, $I[V_{(1)} - \phi_q - d]$, and $I[W_{(1)} - \phi_q + \phi_p - d]$. The above relation can be expressed as

$$\begin{aligned}
&\int_{-\infty}^{\infty}\int_{-\infty}^{\infty}\int_{-\infty}^{\infty}\int_{-\infty}^{\infty}\int_{-\infty}^{\infty}\int_{-\infty}^{\infty} \pi\left(u, U_{(1)}, v, V_{(1)}, w, W_{(1)}\right) \ (uvw)^{N-2} \\
&\times \exp\left(-\left\{\frac{u + NU_{(1)}}{\alpha} + \frac{v + NV_{(1)}}{\beta} + \frac{w + NW_{(1)}}{\gamma}\right\}\right) \\
&\times du\, dU_{(1)}\, dv\, dV_{(1)}\, dw\, dW_{(1)} \;=\; 0, \quad \forall\ \mathbf{\Phi_A}.
\end{aligned} \tag{10.8}$$

The expression on the left hand side of (10.8) is the six-dimensional Laplace transform of the function $\pi(\mathbf{T}')$. It follows from the uniqueness theorem for a two-sided Laplace transform that $\pi(\mathbf{T}') = 0$ almost everywhere, leading to the conclusion that $g(\mathbf{T}') = h(\mathbf{T}')$ and hence there is only one unbiased function of $\mathbf{T}'$. This proves that the statistic $\mathbf{T}'$, or equivalently $\mathbf{T}$, is complete for estimating $\mathbf{\Phi}_A$ when the links are asymmetric and all of α, β, and γ are unknown.

Finally, the complete sufficient statistic $\mathbf{T}$ is also minimal owing to Bahadur's theorem [9], as explained in Chapter 5.

Consequently, finding an unbiased estimator for $\mathbf{\Phi}_A$ as a function of $\mathbf{T}$ yields the MVUE, according to the Rao–Blackwell–Lehmann–Scheffé theorem. Apparently, it is difficult to find six unbiased functions of $\mathbf{T}$ for each of ϕ_q, ϕ_p, d, α, β, and γ just by inspection. But note that ordered BLUE $\hat{\mathbf{\Phi}}_A$ in (10.5) is also an unbiased function of $\mathbf{T}$. Hence, it is concluded that the BLUE based on ordered data is also the MVUE:

$$\hat{\mathbf{\Phi}}_A = \begin{bmatrix} \hat{\phi}_q \\ \hat{\phi}_p \\ \hat{d} \\ \hat{\alpha} \\ \hat{\beta} \\ \hat{\gamma} \end{bmatrix} = \frac{1}{N-1} \begin{bmatrix} N\left(2V_{(1)} - U_{(1)} - W_{(1)}\right) - (2\overline{V} - \overline{U} - \overline{W}) \\ N\left(V_{(1)} - W_{(1)}\right) - (\overline{V} - \overline{W}) \\ N\left(U_{(1)} - V_{(1)} + W_{(1)}\right) - (\overline{U} - \overline{V} + \overline{W}) \\ N\left(\overline{U} - U_{(1)}\right) \\ N\left(\overline{V} - V_{(1)}\right) \\ N\left(\overline{W} - W_{(1)}\right) \end{bmatrix}.$$

The covariance matrix of this estimator is given by (10.3) and hence minimum variances of the clock offsets, fixed and mean delay parameters are given by its diagonal elements, whereas the total MSE for the vector parameter $\hat{\mathbf{\Phi}}_A$ is given by the trace of this matrix.

As a result, the MVUE for the desired parameter, the clock offset of the inactive nodes, for asymmetric unknown network delays is expressed as

$$\hat{\phi}_q = \frac{1}{(N-1)} \left[N\left(2V_{(1)} - U_{(1)} - W_{(1)}\right) - (2\overline{V} - \overline{U} - \overline{W}) \right],$$

and its variance, equal to its MSE, takes the form

$$\text{var}(\hat{\phi}_q) = \frac{1}{N(N-1)} \left(\alpha^2 + 4\beta^2 + \gamma^2\right).$$

10.1.3 Minimum Mean Square Error (MMSE) Estimation

Finding the MMSE estimator is not a straightforward task in any scenario, but Mann [57] described a method to find the optimum estimator for linear functions of the location and scale parameters of a distribution with smallest MSE among estimators with expected loss independent of the location parameters (clock offset and fixed portion of delay in the current problem). Since the derived MMSE estimator is a function of MVUE which was determined in the previous subsection, we can proceed with the derivation of the MMSE estimator expressions for the clock offset, fixed delays, and mean variable delays.

Generalizing the scalar case in [57] to a vector parameter case, for any distribution depending on location and scale parameters only, let $\mathbf{\Delta} \triangleq [\phi_q \; \phi_p \; d]^T$ and $\Xi \triangleq [\alpha \; \beta \; \gamma]^T$. If the unique joint minimum variance unbiased estimator is denoted by $[\hat{\boldsymbol{\Delta}}\ \hat{\boldsymbol{\Xi}}]$, and its covariance matrix is given by (10.3), then the unique joint minimum mean square error with expected loss invariant under transformations of location and scale is

$$\begin{aligned}
\hat{\boldsymbol{\Delta}}^A_{MMSE} &= \hat{\boldsymbol{\Delta}}^A_{MVUE} - \mathbf{E}_2\mathbf{J}^{-1}\left(\mathbf{I} + \mathbf{E}_4\mathbf{J}^{-1}\right)^{-1}\hat{\boldsymbol{\Xi}}^A_{MVUE}, \\
\hat{\boldsymbol{\Xi}}^A_{MMSE} &= \left(\mathbf{I} + \mathbf{E}_4\mathbf{J}^{-1}\right)^{-1}\hat{\boldsymbol{\Xi}}^A_{MVUE},
\end{aligned}$$

where

$$\mathbf{J} = \begin{bmatrix} \alpha^2 & 0 & 0 \\ 0 & \beta^2 & 0 \\ 0 & 0 & \gamma^2 \end{bmatrix}.$$

It is evident that

$$\left(\mathbf{I} + \mathbf{E}_4\mathbf{J}^{-1}\right)^{-1} = \frac{N-1}{N}\,\mathbf{I}.$$

As a result,

$$\begin{aligned}
&\mathbf{E}_2\mathbf{J}^{-1}\left(\mathbf{I} + \mathbf{E}_4\mathbf{J}^{-1}\right)^{-1}\hat{\boldsymbol{\Xi}}^A_{MVUE} \\
&\quad = \frac{1}{N(N-1)}\frac{N-1}{N}\begin{bmatrix} 1 & -2 & 1 \\ 0 & -1 & 1 \\ -1 & 1 & -1 \end{bmatrix}\frac{N}{N-1}\begin{bmatrix} \overline{U} - U_{(1)} \\ \overline{V} - V_{(1)} \\ \overline{W} - W_{(1)} \end{bmatrix} \\
&\quad = \frac{1}{N(N-1)}\begin{bmatrix} \overline{U} - U_{(1)} - 2\left(\overline{V} - V_{(1)}\right) + \overline{W} - W_{(1)} \\ -\left(\overline{V} - V_{(1)}\right) + \overline{W} - W_{(1)} \\ -\left(\overline{U} - U_{(1)}\right) + \overline{V} - V_{(1)} - \left(\overline{W} - W_{(1)}\right) \end{bmatrix}.
\end{aligned}$$

Therefore, the MMSE estimators of the clock offset, fixed delay parameter can be expressed as

$$\begin{aligned}
\hat{\boldsymbol{\Delta}}^A_{MMSE} &= \hat{\boldsymbol{\Delta}}^A_{MVUE} - \mathbf{E}_2\left(\mathbf{I}' + \mathbf{E}_4\right)^{-1}\hat{\boldsymbol{\Xi}}^A_{MVUE} \\
&= \frac{1}{N(N-1)}\begin{bmatrix} (N^2-1)\left(2V_{(1)} - U_{(1)} - W_{(1)}\right) - (N-1)\left(2\overline{V} - \overline{U} - \overline{W}\right) \\ (N^2-1)\left(V_{(1)} - W_{(1)}\right) - (N-1)\left(\overline{V} - \overline{W}\right) \\ (N^2-1)\left(U_{(1)} - V_{(1)} + W_{(1)}\right) - (N-1)\left(2\overline{U} - \overline{V} + \overline{W}\right) \end{bmatrix} \\
&= \frac{1}{N}\begin{bmatrix} (N+1)\left(2V_{(1)} - U_{(1)} - W_{(1)}\right) - \left(2\overline{V} - \overline{U} - \overline{W}\right) \\ (N+1)\left(V_{(1)} - W_{(1)}\right) - \left(\overline{V} - \overline{W}\right) \\ (N+1)\left(U_{(1)} - V_{(1)} + W_{(1)}\right) - \left(\overline{U} - \overline{V} + \overline{W}\right) \end{bmatrix},
\end{aligned}$$

where the MMSE estimator for the mean variable link delays is

$$\hat{\Xi}^{A}_{MMSE} = \left(\mathbf{I} + \mathbf{E}_4\mathbf{J}^{-1}\right)^{-1} \hat{\Xi}^{A}_{MVUE} = \begin{bmatrix} \overline{U} - U_{(1)} \\ \overline{V} - V_{(1)} \\ \overline{W} - W_{(1)} \end{bmatrix}.$$

Therefore, the MMSE estimator for the clock offset of the inactive node assumes the expression

$$\hat{\phi}_q = \frac{1}{N}\left[(N+1)\left(2V_{(1)} - U_{(1)} - W_{(1)}\right) - (2\overline{V} - \overline{U} - \overline{W})\right],$$

and its mean square error is given by

$$\mathrm{MSE}(\hat{\phi}_q) = \frac{N+1}{N^3}\left(\alpha^2 + 4\beta^2 + \gamma^2\right),$$

which clearly outperforms the MVUE.

10.2 Symmetric Exponential Link Delays

The symmetric network delay assumption holds true for some realistic scenarios, e.g., when the nodes have a direct communication link between them and the topology of the network is constant. In this case, $\alpha = \beta = \gamma \triangleq \lambda$.

10.2.1 Best Linear Unbiased Estimation Using Order Statistics (BLUE-OS)

Consider the BLUE based on the ordered data $\mathbf{\Phi}_S \triangleq [\phi_q\ \phi_p\ d\ \lambda]^T$, which is a linear function of an ordered set of observations $\{U_{(k)}\}_{k=1}^N$, $\{V_{(k)}\}_{k=1}^N$, and $\{W_{(k)}\}_{k=1}^N$. Let $\mathbf{z} \triangleq [U_{(1)}\ U_{(2)} \cdots U_{(N)}\ V_{(1)}\ V_{(2)} \cdots V_{(N)}\ W_{(1)}\ W_{(2)} \cdots W_{(N)}]^T$. Then, it is evident that

$$E\left[\mathbf{z}\right] = \begin{bmatrix}\mathbf{A}_1 & \mathbf{A}_2 & \mathbf{A}_3\end{bmatrix}^T \mathbf{\Phi}_S = \mathbf{Q}\mathbf{\Phi}_S,$$

where $\mathbf{z}$ is the $3N \times 1$ ordered data vector, $\mathbf{\Phi}_S$ is the 4×1 vector of unknown parameters and $\mathbf{Q}$ is a known matrix of dimension $3N \times 4$ composed of submatrices

$$\mathbf{A}_1 = \begin{bmatrix} 0 & 0 & \cdots & 0 \\ 1 & 1 & \cdots & 1 \\ 1 & 1 & \cdots & 1 \\ \frac{1}{N} & \frac{1}{N} + \frac{1}{N-1} & \cdots & \sum\limits_{k=1}^{N} \frac{1}{(N-k+1)} \end{bmatrix}, \mathbf{A}_2 = \begin{bmatrix} 1 & 1 & \cdots & 1 \\ 0 & 0 & \cdots & 0 \\ 1 & 1 & \cdots & 1 \\ \frac{1}{N} & \frac{1}{N} + \frac{1}{N-1} & \cdots & \sum\limits_{k=1}^{N} \frac{1}{(N-k+1)} \end{bmatrix},$$

$$\mathbf{A}_3 = \begin{bmatrix} 1 & 1 & \cdots & 1 \\ -1 & -1 & \cdots & -1 \\ 1 & 1 & \cdots & 1 \\ \frac{1}{N} & \frac{1}{N} + \frac{1}{N-1} & \cdots & \sum\limits_{k=1}^{N} \frac{1}{(N-k+1)} \end{bmatrix}.$$

The Gauss–Markov theorem yields the estimator $\hat{\mathbf{\Phi}}_S$ as

$$\hat{\mathbf{\Phi}}_S = \left(\mathbf{Q}^T\mathbf{C}_\mathbf{z}^{-1}\mathbf{Q}\right)^{-1}\mathbf{Q}^T\mathbf{C}_\mathbf{z}^{-1}\mathbf{z}.$$

Since $\{U_{(k)}\}_{k=1}^N$, $\{V_{(k)}\}_{k=1}^N$, and $\{W_{(k)}\}_{k=1}^N$ are independent data sets, $\mathbf{C}_\mathbf{z}$ and its inverse $\mathbf{C}_\mathbf{z}^{-1}$ are now given by

$$\mathbf{C}_\mathbf{z} = \lambda^2 \begin{bmatrix} \mathbf{C} & \mathbf{0} & \mathbf{0} \\ \mathbf{0} & \mathbf{C} & \mathbf{0} \\ \mathbf{0} & \mathbf{0} & \mathbf{C} \end{bmatrix}, \qquad \mathbf{C}_\mathbf{z}^{-1} = \frac{1}{\lambda^2} \begin{bmatrix} \mathbf{C}^{-1} & \mathbf{0} & \mathbf{0} \\ \mathbf{0} & \mathbf{C}^{-1} & \mathbf{0} \\ \mathbf{0} & \mathbf{0} & \mathbf{C}^{-1} \end{bmatrix}.$$

It follows that

$$\mathbf{Q}^T\mathbf{C}_\mathbf{z}^{-1}\mathbf{Q} = \frac{1}{\lambda^2} \begin{bmatrix} 0 & 2N^2 & -N^2 & 0 \\ 2N^2 & -N^2 & 2N^2 & 2N \\ 3N^2 & 0 & 2N^2 & 3N \\ 3N & 0 & 2N & 3N \end{bmatrix},$$

and its inverse is

$$\left(\mathbf{Q}^T\mathbf{C}_\mathbf{z}^{-1}\mathbf{Q}\right)^{-1} = \frac{\lambda^2}{N^2} \begin{bmatrix} 6 & 3 & -4 & 0 \\ 3 & 2 & -2 & 0 \\ -4 & -2 & \frac{9N-8}{3(N-1)} & -\frac{N}{3(N-1)} \\ 0 & 0 & -\frac{N}{3(N-1)} & \frac{N^2}{3(N-1)} \end{bmatrix}. \tag{10.9}$$

As a result,

$$\left(\mathbf{Q}^T\mathbf{C}_\mathbf{z}^{-1}\mathbf{Q}\right)^{-1}\mathbf{Q}^T\mathbf{C}_\mathbf{z}^{-1} = \left[\mathbf{B}_1 \quad \mathbf{B}_2 \quad \mathbf{B}_3\right],$$

where $\mathbf{B}_1$, $\mathbf{B}_2$, and $\mathbf{B}_3$ are defined as

$$\mathbf{B}_1 = \begin{bmatrix} -1 & 0 & \cdots & 0 \\ 0 & 0 & \cdots & 0 \\ 1+\frac{1}{3N} & -\frac{1}{3N(N-1)} & \cdots & -\frac{1}{3N(N-1)} \\ -\frac{1}{3} & \frac{1}{3(N-1)} & \cdots & \frac{1}{3(N-1)} \end{bmatrix},$$

$$\mathbf{B}_2 = \begin{bmatrix} 2 & 0 & \cdots & 0 \\ 1 & 0 & \cdots & 0 \\ \frac{1}{3N}-1 & -\frac{1}{3N(N-1)} & \cdots & -\frac{1}{3N(N-1)} \\ -\frac{1}{3} & \frac{1}{3(N-1)} & \cdots & \frac{1}{3(N-1)} \end{bmatrix},$$

$$\mathbf{B}_3 = \begin{bmatrix} -1 & 0 & \cdots & 0 \\ -1 & 0 & \cdots & 0 \\ 1+\frac{1}{3N} & -\frac{1}{3N(N-1)} & \cdots & -\frac{1}{3N(N-1)} \\ -\frac{1}{3} & \frac{1}{3(N-1)} & \cdots & \frac{1}{3(N-1)} \end{bmatrix}.$$

Therefore, the BLUE using order statistics in the symmetric exponential network delays case is given by

$$\hat{\mathbf{\Phi}}_S = \begin{bmatrix} 2V_{(1)} - U_{(1)} - W_{(1)} \\ V_{(1)} - W_{(1)} \\ \left(1 + \frac{1}{3N}\right)\left(U_{(1)} + W_{(1)}\right) + \left(\frac{1}{3N} - 1\right)V_{(1)} - \frac{1}{3N(N-1)}\sum_{k=2}^{N}\left(U_{(k)} + V_{(k)} + W_{(k)}\right) \\ -\frac{1}{3}\left(U_{(1)} + V_{(1)} + W_{(1)}\right) + \frac{1}{3(N-1)}\sum_{k=2}^{N}\left(U_{(k)} + V_{(k)} + W_{(k)}\right) \end{bmatrix}$$
$$= \begin{bmatrix} 2V_{(1)} - U_{(1)} - W_{(1)} \\ V_{(1)} - W_{(1)} \\ \frac{3N\left(U_{(1)}+W_{(1)}-V_{(1)}\right)+2\left(2V_{(1)}-U_{(1)}-W_{(1)}\right)-\left(\overline{U}+\overline{V}+\overline{W}\right)}{3(N-1)} \\ \frac{N\left\{\left(\overline{U}+\overline{V}+\overline{W}\right)-\left(U_{(1)}+V_{(1)}+W_{(1)}\right)\right\}}{3(N-1)} \end{bmatrix}, \tag{10.10}$$

where $\overline{U}$ and $\overline{V}$ represent the sample averages of the data sets $\{U_k\}_{k=1}^N$ and $\{V_k\}_{k=1}^N$, respectively, and coincide with the sample averages of ordered observations $\{U_{(k)}\}_{k=1}^N$ and $\{V_{(k)}\}_{k=1}^N$. Note that the BLUE of the clock offset based on order statistics matches the MLE of the clock offset in (10.1).

10.2.2 Minimum Variance Unbiased Estimation (MVUE)

In the symmetric case when $\alpha = \beta = \gamma \triangleq \lambda$, the likelihood function for the clock offset as a function of observations $\{U_k\}_{k=1}^N$, $\{V_k\}_{k=1}^N$, and $\{W_k\}_{k=1}^N$ from (9.1), (9.2), and (9.3) can be expressed as

$$\begin{aligned} L\left(\phi_q, \phi_p, d, \lambda\right) &= \lambda^{-3N} \exp\left(-\frac{1}{\lambda}\sum_{k=1}^{N}\left[U_k + V_k + W_k - 2\phi_q - 3d\right]\right) \\ &\quad \times I\left[U_{(1)} - \phi_p - d\right] I\left[V_{(1)} - \phi_q - d\right] I\left[W_{(1)} - \phi_q + \phi_p - d\right]. \end{aligned}$$

Note that due to the equality of the sample means for both raw and ordered observations, the above likelihood function can be factored as

$$\begin{aligned} L\left(\phi_q, \phi_p, d, \lambda\right) &= g_1\left(\sum_{k=1}^{N} U_{(k)}, \sum_{k=1}^{N} V_{(k)}, \sum_{k=1}^{N} W_{(k)}, \lambda\right) g_2\left(U_{(1)}, \phi_p, d\right) \\ &\quad \times g_3\left(V_{(1)}, \phi_q, d\right) g_4\left(W_{(1)}, \phi_q, \phi_p, d\right) h_1\left(\phi_q, d, \lambda\right), \end{aligned}$$

where

$$g_1\left(\sum_{k=1}^{N} U_{(k)}, \sum_{k=1}^{N} V_{(k)}, \sum_{k=1}^{N} W_{(k)}, \lambda\right) = \exp\left(-\frac{1}{\lambda}\sum_{k=1}^{N}\left[U_{(k)} + V_{(k)} + W_{(k)}\right]\right),$$
$$g_2\left(U_{(1)}, \phi_p, d\right) = I\left[U_{(1)} - \phi_p - d\right], \; g_3\left(V_{(1)}, \phi_q, d\right) = I\left[V_{(1)} - \phi_q - d\right],$$
$$g_4\left(W_{(1)}, \phi_q, \phi_p, d\right) = I\left[W_{(1)} - \phi_q + \phi_p - d\right],$$
$$h_1\left(\phi_q, d, \lambda\right) = \lambda^{-3N}\exp\left(\frac{N}{\lambda}\left[2\phi_q + 3d\right]\right).$$

It is evident that $\mathbf{T} = \{\sum_{k=1}^{N}(U_{(k)} + V_{(k)} + W_{(k)}), U_{(1)}, V_{(1)}, W_{(1)}\}$ is the minimal sufficient statistic according to Neymann–Fisher factorization theorem. Now proceeding similarly as before, $\sum_{k=1}^{N}(U_{(k)} + V_{(k)} + W_{(k)})$ is dependent on $U_{(1)}$, $V_{(1)}$, and $W_{(1)}$. As a result, $\mathbf{T}$ can be transformed into $\mathbf{T}' = \{\sum_{k=1}^{N}(U_{(k)} - U_{(1)} + V_{(k)} - V_{(1)} + W_{(k)} - W_{(1)}), U_{(1)}, V_{(1)}, W_{(1)}\}$. It can be concluded from the discussion in the last section that $\sum_{k=1}^{N}(U_{(k)} - U_{(1)} + V_{(k)} - V_{(1)} + W_{(k)} - W_{(1)})$ is Gamma distributed with parameters $(3(N-1), \lambda)$. Hence, $\mathbf{T}'$ is a combination of four independent random variables, which in terms of the three-parameter Gamma distribution are

$$q = \sum_{k=1}^{N}(U_{(k)} - U_{(1)} + V_{(k)} - V_{(1)} + W_{(k)} - W_{(1)}) \sim \Gamma\left(3\left(N-1\right), \lambda, 0\right),$$
$$U_{(1)} \sim \Gamma\left(1, \frac{\lambda}{N}, \phi_p + d\right), V_{(1)} \sim \Gamma\left(1, \frac{\lambda}{N}, \phi_q + d\right), W_{(1)} \sim \Gamma\left(1, \frac{\lambda}{N}, \phi_q - \phi_p + d\right).$$

Next, defining $g(\mathbf{T}')$ and $h(\mathbf{T}')$ as functions of $\mathbf{T}'$ such that $E[g(\mathbf{T}')] = E[h(\mathbf{T}')] = \mathbf{\Phi}_S$,

$$E\left[g\left(\mathbf{T}'\right) - h\left(\mathbf{T}'\right)\right] = E\left[\pi\left(\mathbf{T}'\right)\right] = 0, \qquad \forall\ \mathbf{\Phi}_S,$$

where the expectation is taken with respect to $p(\mathbf{T}'; \mathbf{\Phi}_S)$. As a result, since the domain of q is also dictated by $U_{(1)}$, $V_{(1)}$, and $W_{(1)}$,

$$\begin{aligned}\iiiint\limits_{R_{U_{(1)},V_{(1)},W_{(1)}}} &\pi\left(q, U_{(1)}, V_{(1)}, W_{(1)}\right) \frac{\lambda^{-3(N-1)}}{\Gamma\left[3\left(N-1\right)\right]} q^{3N-4}\exp\left(-\frac{q}{\lambda}\right)\left(\frac{N}{\lambda}\right)^3 \\ &\times \exp\left(-\frac{N}{\lambda}\left\{U_{(1)} + V_{(1)} + W_{(1)} - 2\phi_q - 3d\right\}\right) \\ &\times dq\, dU_{(1)}\, dV_{(1)}\, dW_{(1)} = 0, \qquad \forall\ \mathbf{\Phi}_S,\end{aligned}$$

where $R_{U_{(1)},V_{(1)},W_{(1)}}$ is the region defined by $I[U_{(1)} - \phi_p - d]$, $I[V_{(1)} - \phi_q - d]$, and $I[W_{(1)} - \phi_q + \phi_p - d]$. It follows that

$$\begin{aligned}\int_{-\infty}^{\infty}\int_{-\infty}^{\infty}\int_{-\infty}^{\infty}\int_{-\infty}^{\infty} \pi\left(q, U_{(1)}, V_{(1)}, W_{(1)}\right) q^{3N-4} \\ \times \exp\left(-\frac{N}{\lambda}\left\{\frac{q}{N} + U_{(1)} + V_{(1)} + W_{(1)}\right\}\right) \\ \times\, dq\, dU_{(1)}\, dV_{(1)}\, dW_{(1)} &= 0, \qquad \forall\ \mathbf{\Phi}_S.\end{aligned}$$

From the uniqueness theorem for the two-sided Laplace transform, it follows that $\pi(\mathbf{T}') = 0$ almost everywhere, resulting in the completeness of $\mathbf{T}'$, or equivalently $\mathbf{T}$. Hence, $\mathbf{T}$ is also the minimal sufficient statistics from Bahadur's theorem and the MVUE is the same as $\hat{\mathbf{\Phi}}_S$ in (10.10) expressed as

$$\hat{\mathbf{\Phi}}_S = \begin{bmatrix} 2V_{(1)} - U_{(1)} - W_{(1)} \\ V_{(1)} - W_{(1)} \\ \frac{1}{3(N-1)}\left\{3N\left(U_{(1)} + W_{(1)} - V_{(1)}\right) + 2\left(2V_{(1)} - U_{(1)} - W_{(1)}\right) - (\overline{U} + \overline{V} + \overline{W})\right\} \\ \frac{N}{3(N-1)}\left\{(\overline{U} + \overline{V} + \overline{W}) - \left(U_{(1)} + V_{(1)} + W_{(1)}\right)\right\} \end{bmatrix}.$$

The covariance matrix of this estimator is given by (10.9) and the diagonal elements represent the variance of each individual unknown parameter, whereas the trace of this matrix is the total MSE or variance for the vector parameter $\mathbf{\Phi}_S$.

Hence, the MVUE for the clock offset of the inactive node in the case of symmetric unknown network delays is expressed as

$$\hat{\phi}_q = 2V_{(1)} - U_{(1)} - W_{(1)},$$

and its variance takes the form

$$\operatorname{var}\left(\hat{\phi}_q\right) = \frac{6\lambda^2}{N^2}.$$

10.2.3 Minimum Mean Square Error (MMSE) Estimation

Proceeding similarly as before, let $\mathbf{\Delta}^S_{MMSE} \triangleq [\phi_q\ \phi_p\ d]^T$ and $\Xi^S_{MMSE} \triangleq \lambda$. If the unique joint minimum variance unbiased estimator is denoted $[\hat{\mathbf{\Delta}}^S_{MVUE}\ \hat{\Xi}^S_{MVUE}]$, and with the covariance matrix (10.9) given by

$$\begin{bmatrix} \mathbf{E}_1 & \mathbf{E}_2 \\ \mathbf{E}_3 & \mathbf{E}_4 \end{bmatrix},$$

where

$$\mathbf{E}_1 = \frac{\lambda^2}{N^2}\begin{bmatrix} 6 & 3 & -4 \\ 3 & 2 & -2 \\ -4 & -2 & \frac{9N-8}{3(N-1)} \end{bmatrix}, \mathbf{E}_2 = \frac{\lambda^2}{N^2}\begin{bmatrix} 0 \\ 0 \\ -\frac{N}{3(N-1)} \end{bmatrix},$$

$$\mathbf{E}_3 = \frac{\lambda^2}{N^2}\begin{bmatrix} 0 & 0 & -\frac{N}{3(N-1)} \end{bmatrix}, \quad \mathbf{E}_4 = \frac{\lambda^2}{3(N-1)},$$

then the unique joint MMSE estimator with expected loss invariant under transformations of location and scale is given by

$$\begin{aligned} \hat{\boldsymbol{\Delta}}^S_{MMSE} &= \hat{\boldsymbol{\Delta}}^S_{MVUE} - \frac{1}{\lambda^2}\mathbf{E}_2\left(1 + \frac{1}{\lambda^2}\mathbf{E}_4\right)^{-1}\hat{\boldsymbol{\Xi}}^S_{MVUE}, \\ &= \begin{bmatrix} 2V_{(1)} - U_{(1)} - W_{(1)} \\ V_{(1)} - W_{(1)} \\ \hat{d}^S_{MVUE} + \frac{1}{3(N-1)(3N-2)}\hat{\lambda}^S_{MVUE} \end{bmatrix}, \end{aligned}$$

and

$$\begin{aligned} \hat{\boldsymbol{\Xi}}^S_{MMSE} &= \left(1 + \frac{1}{\lambda^2}\mathbf{E}_4\right)^{-1}\hat{\boldsymbol{\Xi}}^S_{MVUE} \\ &= \frac{N}{3N-2}\left[(\overline{U} + \overline{V} + \overline{W}) - \left(U_{(1)} + V_{(1)} + W_{(1)}\right)\right]. \end{aligned}$$

It is evident from the above expressions that the MMSE estimator for the desired clock offset parameter of the inactive nodes is the same as the MLE, ordered BLUE, and the MVUE for the symmetric exponential link delay model.

Chapter 11

ADAPTIVE MULTI-HOP TIME SYNCHRONIZATION (AMTS)

Developing long-term and network-wide timing-synchronization protocols that are energy-efficient represents one of the key strategies for the successful deployment of long-lived sensor networks. However, most of the existing protocols have focused only on achieving synchronization for short timescales, and are not appropriate for long-term synchronization. In the adaptive-clock synchronization protocols [28] and [69], optimizing the network synchronization protocol was considered with the aim of achieving a specific synchronization accuracy with minimal energy consumption. The adaptive-clock synchronization protocol [69] represents a probabilistic extension of RBS and proposes a mechanism for determining the minimum number of synchronization beacons and the synchronization rate in order to achieve a pre-established clock synchronization error. Ganeriwal *et al.* [28] proposed for the first time a measurement-based study for designing an energy-efficient rate-adaptive long-time synchronization protocol (RATS) that adapts the synchronization period, number of beacons, and length of prediction window to achieve an application-specific accuracy.

Motivated in part by these preliminary contributions, we propose a more powerful AMTS scheme with the goal of achieving a long-term network-wide synchronization with minimal energy consumption. AMTS exhibits a number of attractive features:

- It represents a significantly enhanced extension of TPSN aiming at minimizing the overall energy consumption in large-scale and long-lived sensor networks.
- It is equipped with flexible mechanisms to adjust the synchronization mode, the period of network-wide timing synchronization (resynchronization rate),

and schemes for joint estimation of clock offset and skew in order to achieve long-term reliability of synchronization.

- It employs a sequential message exchange technique and an energy-efficient signaling scheme to further reduce the energy consumption in synchronization procedures.

- As opposed to RBS [22] and FTSP [58] that perform very poorly in high-latency acoustic networks, AMTS provides excellent performance in underwater acoustic networks characterized by high propagation delays and possible clock skew variations.

11.1 MAIN IDEAS

AMTS is based on a similar system model to TPSN and employs a number of novel features as well. It consists of three functional phases: the *network level discovery phase*, the *synchronization phase*, and the *network evaluation phase*, and a number of network parameters such as the latency factor, the average number of hops, and the resynchronization period to optimize the synchronization protocol. Relative to TPSN, AMTS assumes the additional *network evaluation phase*, while the functions of the other two phases are similar to the ones encountered in TPSN.

Robustness to high latencies and network delays is ensured based on the clock estimators presented in Chapter 6, and therefore AMTS works well for sensor network applications presenting large delays in timing message exchanges such as underwater acoustic sensor networks [93]. Moreover, AMTS adapts the schemes used for joint estimation of clock offset and skew to increase the resynchronization period. In addition, novel sequential message exchange and an efficient signaling technique are adopted in order to further decrease the energy consumption.

As with TPSN, generating a hierarchical structure in the network, the level discovery phase, is the first step of AMTS. In this phase, every single node in the network is assigned a level and is prepared for synchronization. The second step of AMTS, called the time synchronization phase, consists of pairwise synchronizations between adjacent nodes until every node in the network is synchronized to the reference. In the synchronization phase, AMTS estimates not only the current clock offset but also the clock frequency (skew) to guarantee long-term reliability of synchronization while TPSN only estimates the clock offset. Hence, AMTS will be expected to require far less frequent resynchronization than TPSN. Finally, the reference node investigates the current status of network traffic in order to optimize the resynchronization period and the number of beacons in terms of energy efficiency. In addition, AMTS selects the synchronization mode between the *always on*

(AO) mode (always maintain network-wide synchronization) and the *sensor initiated* mode (SI) (synchronize only when it needs to) based on the network status. This step is the network evaluation phase, and its goal is to minimize the number of message exchanges for synchronization in a given time, i.e., it aims to minimize the total energy consumption of synchronization. AMTS periodically repeats the synchronization and network evaluation phases to minimize the total energy consumption with respect to the current network status. The functional phases of AMTS are summarized below (with the second and third phases being repeated periodically).

- *Level discovery phase* This is the same as that in TPSN, and is used for generating a hierarchical structure in the network.
- *Synchronization phase* This is similar to the corresponding synchronization phase in TPSN. However, AMTS adjusts not only the current clock offset but also the clock skew to guarantee the long-term synchronization, while TPSN only estimates the clock offset. Hence, AMTS requires far less frequent resynchronization relative to TPSN.
- *Network evaluation phase* The reference node investigates the current status of network traffic in order to select the synchronization mode between the AO mode (always maintain network-wide synchronization) and the SI mode (synchronize only when it needs to). In addition, it optimizes the resynchronization period and the number of beacons per each pairwise synchronization.

The following sections describe these phases in detail.

11.2 LEVEL DISCOVERY PHASE

As in TPSN, the role of the level discovery phase in AMTS is to create a hierarchical structure (spanning tree) of the network. The level discovery phase in AMTS consists of the following steps: 1. Select a root node using an appropriate leader election algorithm and assign a zero level to the root node. 2. Broadcast a level discovery packet (LDP) containing the identity and the level of packet. 3. Assign a level that is one greater than that of the received packet to every node that receives an LDP which sends a new LDP attaching it own level. 4. Repeat this process until every node in the network is successfully assigned a level. After being assigned a level, every node discards further packets requesting level discovery to prevent collisions.

11.3 SYNCHRONIZATION PHASE

This phase performs pairwise synchronization between a set of nodes by exchanging timing messages. For the AO mode, a series of pairwise synchronizations take place until every node in the network is synchronized to the reference, i.e., the message exchanges are occurring at all branches of the network spanning tree. On the other hand, for the SI mode, only the nodes participating in the particular multi-hop data transmission synchronize with each other. AMTS adapts the joint clock offset and skew estimation mechanism proposed in Chapter 6 by considering the long-term reliability and energy efficiency of synchronization as design criteria.

11.4 NETWORK EVALUATION PHASE

In this phase, the network examines the total number of message exchanges required for synchronization during the last synchronization period, then adjusts the duration of the next synchronization period to minimize the overall energy consumption for synchronization. When the network traffic occurs rarely and synchronization delay is not a critical problem, applying the SI mode is a better choice for saving network resources instead of using the AO mode. In addition, for some applications, the sensor clocks might be allowed to go out of synchronization unless sensing events happen. Another critical problem is to determine the required number of timing message exchanges (beacons) per each pairwise synchronization. For synchronization accuracy, a large number of message transfers and corresponding signal processing is needed in each pairwise synchronization. However, as the number of required timing messages per each pairwise synchronization increases, the overall number of timing messages in a synchronization period increases. Hence, there is a tradeoff between accuracy and energy consumption.

11.4.1 SYNCHRONIZATION MODE SELECTION

To address the above design challenges, AMTS determines the following network parameters: the synchronization mode (AO or SI), the resynchronization period τ, and the number of beacons per pairwise synchronization N. AMTS aims at efficient usage of network resources (i.e., energy saving) in synchronization. The idea of selecting the synchronization mode between AO and SI is based on the observation that when the network traffic occurs rarely and synchronization delay is not a critical problem, keeping all the sensor nodes synchronized all the times (AO mode) is not a good strategy since synchronization consumes a lot of energy. In addition, for some applications, the sensor clocks might be allowed to go out of synchronization unless sensing events happen. In this case, the SI mode, in which only nodes participating in a particular multi-hop data transmission synchronize with each other, is a better choice. We define the following parameters:

- B The number of branches (edges) in a spanning tree of the network, which can be obtained after the level discovery phase.

- τ The (re)synchronization period, i.e., the time between resynchronization.
- $\bar{h}$ The average number of hops per unit time. In every sensing event, the destination node records the number of hops that occurred in that transmission. During the synchronization phase, the reference node collects the information about the total number of hops that occurred in the last synchronization period and determines the average number of hops per unit time ($\bar{h}$) in the network. This information indicates how busy the network traffic is and can be included in the timing messages with a small overhead.
- δ The latency factor ($0 \leq \delta \leq 1$) reflecting the amount of allowed delay in data transmission. A high latency factor means less concern for network delays. For example, δ is set to be 0 for sensor networks that require network synchronization all the time. At the other extreme, for delay-independent networks, δ should be close to 1.
- N The number of timing message exchanges per pairwise synchronization.

As mentioned before, the goal of AMTS is to minimize the number of required timing messages. In the AO mode, the number of timing messages per unit time is given by $\overline{M} = 2BN/\tau$, while in the SI mode $\overline{M} = 2\bar{h}N$. To minimize the number of timing messages per unit time $\overline{M}$, the synchronization mode should be selected as follows:

$$\frac{2BN\delta}{\tau} \underset{\text{SI}}{\overset{\text{AO}}{\lessgtr}} 2\bar{h}N, \tag{11.1}$$

where the latency factor δ takes values between 0 and 1 such that the more delay-dependent networks assume a larger value of δ and vise versa ($0 \leq \delta \leq 1$).

As the clock synchronization period τ increases, the network becomes more power efficient. Thus, τ should be chosen as large as possible. However, too large a value of τ induces a critical synchronization problem since the clock difference (offset) between nodes generally keeps increasing with time. Hence, there exists a maximum time synchronization period (τ_{max}) which is determined by the oscillator characteristics (hardware specifications) and the accuracy of estimators. Notice that sensing data transmission is not available during the synchronization phase (τ_{sync}), therefore the resynchronization period is given by $\tau = \tau_{max} + \tau_{sync}$. In the sequel, (11.1) is expressed as

$$\tau \underset{\text{SI}}{\overset{\text{AO}}{\gtrless}} \frac{B\delta}{\bar{h}}. \tag{11.2}$$

From (11.2), the synchronization mode changes from AO to SI when τ is smaller than $B\delta/\bar{h}$ and vice versa. In the SI mode, the reference node periodically asks how many hops occurred during the previous time interval, and then makes a decision whether or not to switch to the AO mode. Actually, τ is also dependent on N since it strongly depends on the accuracy of timing offset estimators. A more detailed analysis of τ is provided in the following subsection.

11.4.2 DETERMINATION OF SYNCHRONIZATION PERIOD

As mentioned in the previous subsection the resynchronization period (τ_{max}) should be determined by the accuracy of the estimators and the oscillator specifications. A large value of τ_{max} is desirable in order to have a more power-efficient network. However, a reduced value of τ_{max} is necessary because the phase offset between the nodes increases as time progresses.

Suppose that the clock timing mismatch ε between the two nodes is modeled as follows: $\varepsilon = \varepsilon_o + \varepsilon_s t$, where t denotes the reference time, ε_o and ε_s stand for the clock offset and skew errors, respectively. Let $\varepsilon_{o,i}$ and $\varepsilon_{s,i}$ denote the clock offset and skew estimation errors when i messages are exchanged between the two nodes. In general, it is difficult to determine a specific mathematical model for either the clock offset or skew errors. Here, we model both clock offset and skew errors by normal distributions in accordance with the experimental results reported in [22] and [29]:

$$\varepsilon_{o,i} \sim \mathcal{N}(0, \sigma^2_{\varepsilon_{o,i}}) \quad 1 \leq i \leq N,$$
$$\varepsilon_{s,i} \sim \mathcal{N}(0, \sigma^2_{\varepsilon_{s,i}}) \quad 1 \leq i \leq N.$$

Note that clock skew estimation is only available when there are multiple message exchanges. Hence, $\varepsilon_{s,1}$ stands for the clock skew error when there is no skew estimation. Here, the maximum clock mismatch can be modeled as another normal distribution $\varepsilon \sim \mathcal{N}(0, \sigma^2_\varepsilon)$, where $\sigma^2_\varepsilon = \sigma^2_{\varepsilon_{o,N}} + \sigma^2_{\varepsilon_{s,N}} \tau^2_{\max}$, $(t = \tau_{max})$. Let $\varepsilon_{\max}$ denote the upper limit for the clock error via the probabilistic measure:

$$P_s = Pr\left(|\varepsilon| \geq \varepsilon_{\max}\right) = \operatorname{erfc}\left(\frac{\varepsilon_{max}}{\sqrt{2}\sigma_\varepsilon}\right),$$

where $\operatorname{erfc}(x) \triangleq (2/\sqrt{\pi}) \int_x^\infty \exp(-t^2)dt$ and P_s denotes the synchronization error probability for pairwise synchronization. Thus, σ_ε can be determined when ε_{max} and the maximum allowable P_s are specified. For instance, when P_s is limited to 0.1% and ε_{max} is 10 ms, then the standard deviation of clock mismatch (σ_ε) has to be smaller than 3.04 ms.

The maximum resynchronization period with N beacons can be expressed as

$$\tau_{max}^{(N)} = \sqrt{\frac{\sigma_{\varepsilon}^2 - \sigma_{\varepsilon_{o,N}}^2}{\sigma_{\varepsilon_{s,N}}^2}}. \tag{11.3}$$

Based on the lower bounds and asymptotic performance of the estimators, one can easily infer closed-form expressions of the variances $\sigma_{\varepsilon_{o,N}}^2$ and $\sigma_{\varepsilon_{s,N}}^2$ in terms of the variances $\sigma_{\varepsilon_{o,1}}^2$ and $\sigma_{\varepsilon_{s,2}}^2$, respectively. From the lower bound derived in Chapter 6 (equation (6.3)), $\sigma_{\varepsilon_{o,N}}^2$ can be expressed with respect to N and $\sigma_{\varepsilon_{o,1}}^2$ as follows:

$$\sigma_{\varepsilon_{o,N}}^2 = \frac{\sigma_{\varepsilon_{o,1}}^2}{N}.$$

Similarly, since the time differences between beacons are proportional to N and by far greater than the variance of delays, the following relationship can be obtained:

$$\sigma_{\varepsilon_{s,N}}^2 = \frac{\sigma_{\varepsilon_{s,2}}^2}{(N-1)^2}, \quad N \geq 2.$$

Therefore, for $N \geq 2$, $\tau_{max}^{(N)}$ can be rewritten as

$$\tau_{max}^{(N)} = (N-1)\sqrt{\frac{\sigma_{\varepsilon}^2 - \frac{\sigma_{\varepsilon_{o,1}}^2}{N}}{\sigma_{\varepsilon_{s,2}}^2}}, \qquad N \geq 2. \tag{11.4}$$

Note that $\varepsilon_{s,1}$ can be determined from the specifications of the crystal oscillator, while $\varepsilon_{o,1}$ and $\varepsilon_{s,2}$ can be determined through simple experimental tests. Therefore, the maximum resynchronization period is proportional to the number of beacons, and performing clock skew estimation will significantly increase $\tau_{max}^{(N)}$ since $\sigma_{\varepsilon_{s,1}} \gg \sigma_{\varepsilon_{s,2}}$.

Let us consider an example. Assume that the upper limit for the clock error ε_{max} is 10 ms, the worst case of synchronization error (ε_o) is 50 μs, and the worst case of clock skew (ε_s) is 4.75 μs/s as used in [29]. Then the maximum timing synchronization period τ_{max} can from obtained from

$$10 \text{ ms} = 50\ \mu\text{s} + 4.75 \times \tau_{max}; \quad \tau_{max} \approx 35 \text{ min}.$$

If these bounds are 99.99% satisfied, the set of standard deviations can be calculated as: $\sigma_{\varepsilon} = 3.33$ ms, $\sigma_{\varepsilon_o} = 16.67$ μs, and $\sigma_{\varepsilon_s} = 1.58$ μs. Then, plugging these values into (11.3) gives $\tau_{max} \approx 35$ min, which matches well with the above result. For $N \geq 2$, the maximum timing synchronization period $\tau_{max}^{(N)}$ becomes close to $35(N-1)\left(\sigma_{\varepsilon,1}/\sigma_{\varepsilon,2}\right)$ min in this example.

11.4.3 DETERMINATION OF THE NUMBER OF BEACONS

The goal of AMTS is to minimize the average number of message exchanges ($\overline{M}$). Hence, from (11.4), finding the optimal number of beacons (N) reduces to solving the following optimization problem:

$$\hat{N} = \arg\min_{N} \overline{M}, \tag{11.5}$$

with

$$\overline{M} = \frac{2BN}{\tau_{sync}^{(N)} + \tau_{\max}^{(N)}} = \begin{cases} \dfrac{2B}{\tau_{sync}^{(1)} + \sqrt{\dfrac{\sigma_{\varepsilon}^2 - \sigma_{\varepsilon_{o,1}}^2}{\sigma_{\varepsilon_{s,1}}^2}}} & N = 1 \\ \dfrac{2B}{\dfrac{\tau_{sync}^{(N)}}{N} + \dfrac{N-1}{N}\sqrt{\dfrac{\sigma_{\varepsilon}^2 - \frac{\sigma_{\varepsilon_{o,1}}^2}{N}}{\sigma_{\varepsilon_{s,2}}^2}}} & N \geq 2 \end{cases},$$

where $\tau_{sync}^{(N)}$ denotes the synchronization time with N beacons and will be estimated at the reference node for different values of N when the network is first established. Once N is estimated from (11.5), $\tau_{max}^{(N)}$ can be obtained from (11.4). Figure 11.1 illustrates the flowchart of the proposed synchronization scheme.

11.4.4 SEQUENTIAL MULTI-HOP SYNCHRONIZATION ALGORITHM (SMA)

This section proposes the SMA for energy-efficient timing message exchanges. The key idea of SMA is that the upper and lower level nodes are able to receive the same timing message simultaneously. The aim of SMA is to reduce the number of timing message exchanges by using this property, and hence reduce the energy consumption for synchronization. Figure 11.2 compares the signaling strategy corresponding to the SMA (Figure 11.2(a)) with the conventional one (Figure 11.2(b)). The assumed network topology is a simple one-dimensional (linear) network with four levels. In TPSN, the clock synchronization must be done step-by-step as depicted in Figure 11.2(b).

In Figure 11.2(a), a node in the third level (node 3) transmits a synchronization packet containing a time stamp (T_1) and a node identifier (ID) to a node in the second level (node 2), then node 2 sends back an acknowledgement packet (ACK) to node 3 containing time stamps (T_1, T_2, and T_3) and its ID. Since a node in the first level (node 1) also receives the ACK from node 2, it sends back an ACK

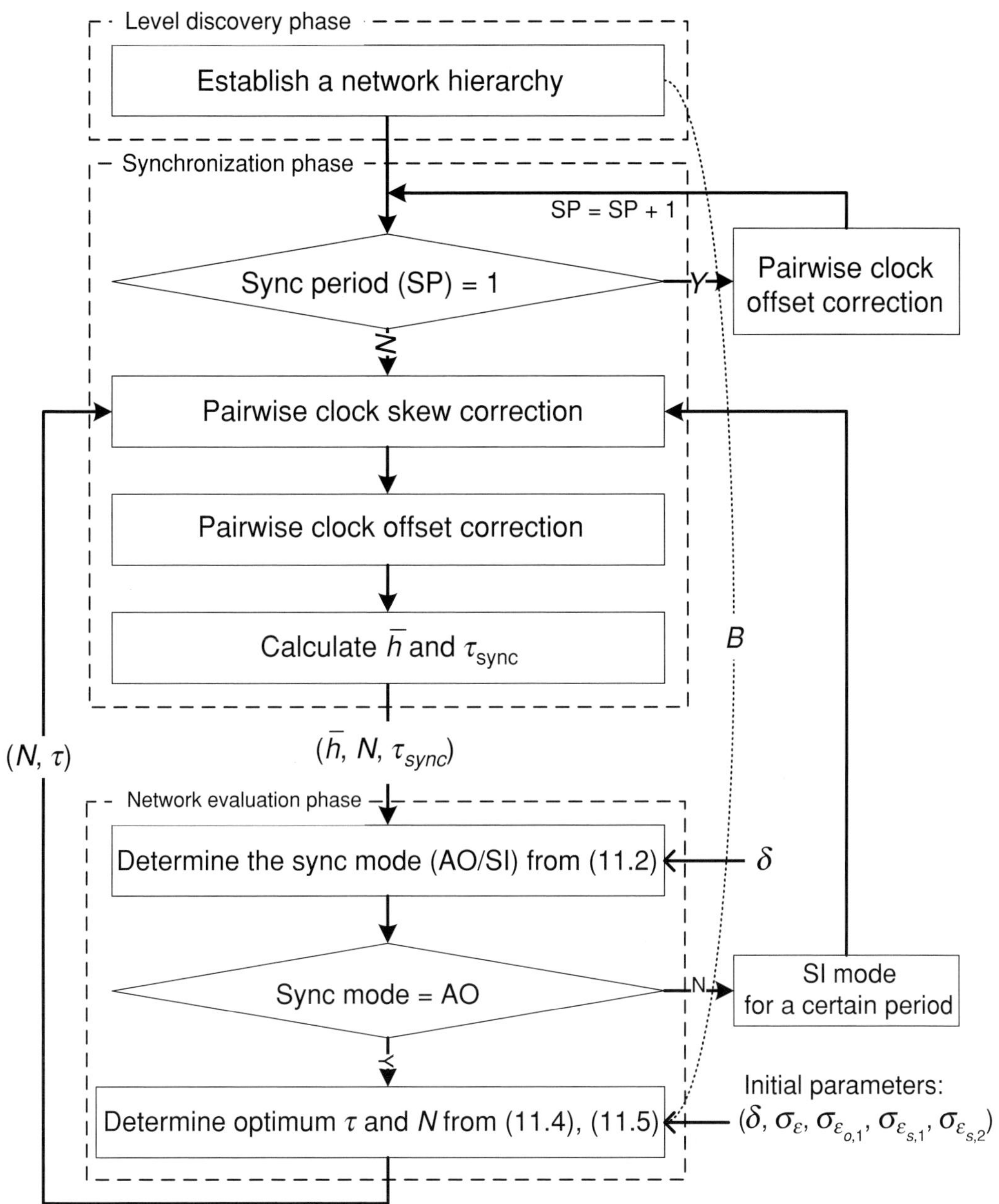

Figure 11.1: Flowchart of AMTS.

having time stamps (T_3, T_5, and T_6) and its ID to node 2. Thus, node 1 regards the ACK from node 2 as a synchronization packet from node 2. Similarly, the reference node (*Node R*) receives the ACK from node 1, then returns an ACK to node 1. This sequential procedure is continued N times so as to increase the accuracy of clock

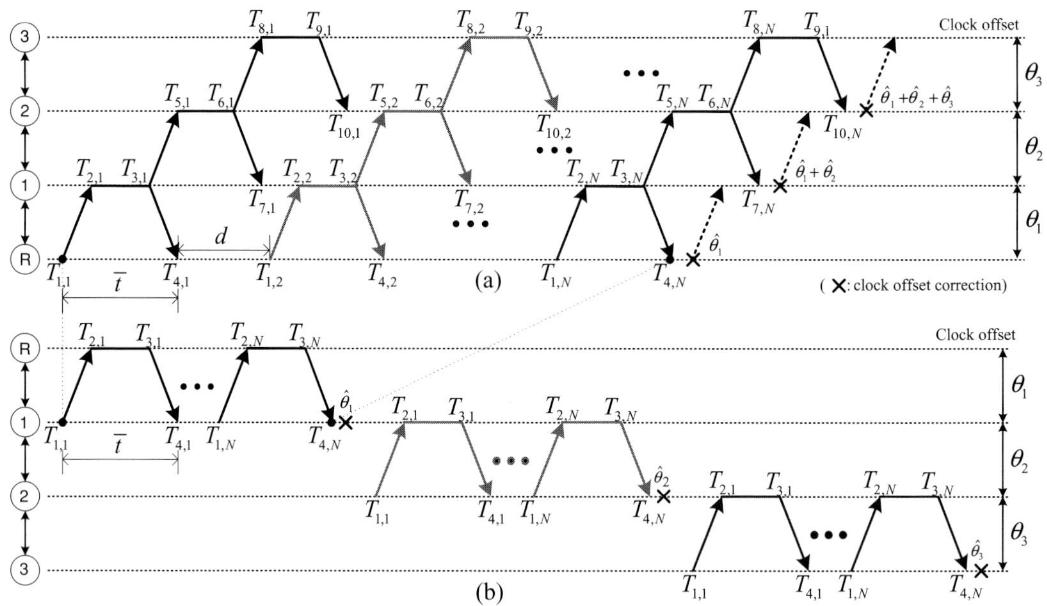

Figure 11.2: Time synchronization models for multi-hop synchronization: (a) SMA; (b) TPSN.

estimation to an adequate level. Finally, node 1 estimates and adjusts the clock offset, then sends it to node 2 to assist node 2 to synchronize to the reference. Node 2 repeats the same procedure so that every node in the network can be synchronized. This procedure can be easily extended to networks with a large number of levels.

SMA presents two major advantages when compared with the signaling scheme adopted in TPSN: a decrease in the number of required timing messages for network-wide synchronization (M) by a factor of 2, and a significant reduction of the synchronization time ($\tau_{sync}^{(N)}$). In TPSN, the number of timing messages M is given by $M = 2BN$, while in SMA the number of timing messages is given by $M = (B+1)N + B$. Hence, the ratio between the number of timing message exchanges required by TPSN and SMA is

$$R_m = \frac{2}{1 + \frac{1}{N} + \frac{1}{B}}. \tag{11.6}$$

Thus, SMA requires fewer timing messages for synchronization than TPSN when there are multiple branches and beacons in the network. In addition, as B and N are sufficiently large, R_m tends to 2.

From Figure 11.2, the synchronization times for SMA and TPSN can be expressed, respectively, as

$$\tau_{sync}^{(N)} \triangleq N(\bar{t}+d) + B\bar{t} \qquad \text{(SMA)}, \tag{11.7}$$

$$\tau_{sync}^{(N)} \triangleq N(\bar{t}+d)B + (B-1)d_L \qquad \text{(TPSN)}, \tag{11.8}$$

where $\bar{t}$ stands for the average time for a timing message exchange (see Figure 11.2), and d and d_L stand for the average delays between beacons ($d \ll d_L$) for SMA and TPSN, respectively. Note that, from (11.7) and (11.8), the synchronization time $\tau_{sync}^{(N)}$ is smaller than $N\tau_{sync}^{(1)}$, and $\tau_{sync}^{(N)}/N$ monotonically decreases as N increases. The ratio between the synchronization times for TPSN and SMA is

$$R_t = \frac{N(\bar{t}+d)B + (B-1)d_L}{N(\bar{t}+d) + B\bar{t}} = \frac{B\left(1 + \frac{d_L}{N(\bar{t}+d)} - \frac{d_L}{BN(\bar{t}+d)}\right)}{\left(1 + \frac{B\bar{t}}{N(\bar{t}+d)}\right)}. \tag{11.9}$$

Thus, if N is sufficiently large, R_t tends to B, which indicates a reduction of synchronization time by a factor of B. Moreover, as the number of branches B increases, R_t converges to $N(1 + d/\bar{t} + d_L/N\bar{t}) \approx N$, and therefore there is a reduction of the synchronization time by a factor of N. For arbitrary two-dimensional networks, the timing message transmissions of the descendent nodes can be scheduled to avoid possible data collisions (which induce additional delays) using transmission scheduling schemes. Developing efficient signaling methods to minimize data collisions and delays in general two-dimensional networks represents an interesting open research problem.

11.5 SIMULATION RESULTS

In Figure 11.3 the performances of AMTS and TPSN are compared in terms of the average number of message exchanges ($\overline{M}$) with respect to the number of beacons N when P_s is 1%, 0.1%, 0.01%, and 0.001%. This simulation is based on the linear network model where the depth of the network $B = 5$, $\varepsilon_{max} = 10\ ms$, $\sigma_{\varepsilon_o} = 16.67$ μs, $d = 10$ ms, $\bar{t} = 400$ ms, and $\sigma_{\varepsilon_s} = 1.58$ μs.

It can be seen that AMTS requires far fewer timing messages than TPSN when multiple beacon transmissions are used. Moreover, the gap between the average number of timing messages required for synchronization in AMTS and TPSN significantly increases as N increases, and thus AMTS is far more energy-efficient than TPSN for large Ns. It can also be seen that just a few beacons is enough to minimize $\overline{M}$ for AMTS. Moreover, as expected, a larger number of beacons is required to meet

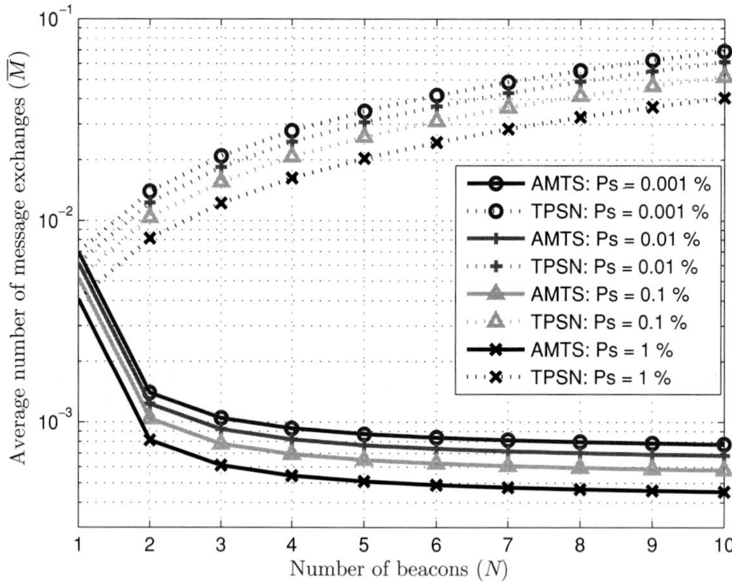

Figure 11.3: Average number of message exchanges ($\overline{M}$) with respect to the number of beacons.

a more strict constraint on the network-wide error probability P_s. In practice, a low N is highly preferable, since N is proportional to the synchronization time, i.e., a low N induces better latency performance (although it may not be optimal in terms of energy consumption). In this simulation, the optimum values of N and $\tau^{(N)}$ are obtained with the constraint $\tau^{(N)} \leq \tau_{sync}^{max}$.

Chapter 12

CLOCK DRIFT ESTIMATION FOR ACHIEVING LONG-TERM SYNCHRONIZATION

Up to now, various schemes have been proposed for estimating the clock offset and skew. However, estimating the clock of a node using a linear model is useful only for short-term applications, examples of which are object tracking and surveillance. It is not sufficient for certain applications with stringent and long-term clock synchronization requirements, such as efficient duty cycling and synchronized sampling, because they spend a lot of energy on resynchronization during a given time interval.

To elaborate on this point, consider the following examples. In FTSP the nodes in the network have to be resynchronized *every minute* to achieve a 90 μs synchronization error, even though it is the most efficient time synchronization protocol reported thus far and has been implemented on real testbeds with very good results [58]. In addition, the Center for Embedded Networked Sensing (CENS) deployment at James Reserves [15] uses RBS to synchronize the nodes after every 5 minutes and the shooter localization system [87] implements FTSP to synchronize once every 45 seconds. Due to the difficulties associated with long-term synchronization, although RBS and FTSP estimate the clock skew alongside clock offset using linear regression, they are not adequate in practice to achieve long-term synchronization since they are confined to estimating only the first-order parameter (clock skew). Hence, to achieving the goal of long-term synchronization, a better modeling of the relationship between the clock and the reference node is required. In this chapter, this problem is targeted through extending the linear model between two clocks to a quadratic one and then the clock parameters of clock offset, skew, and drift are jointly estimated.

Estimating this second-order parameter of the model, called clock drift, is important in light of the reasons mentioned above and finding the MLE is attractive due to its optimal properties for a large number of observations as mentioned before. It should be noted here that although the estimation of clock parameters using a quadratic model is computationally more demanding than that using the linear model, it helps to maintain long-term synchronization among the nodes and consequently there are fewer communications and a resultant power saving. As discussed before, the energy required to transmit 1 bit over 100 meters (3 joules) is roughly equivalent to the energy required to execute 3 *million* of instructions [75]. Therefore the higher computational complexity (and implicitly the increased synchronization accuracy) induced by the estimation of additional parameters such as skew and drift might lead to a reduction in the number of RF transmissions required to maintain the network synchronization. Hence important energy savings might be possible.

12.1 PROBLEM FORMULATION

Now consider a two-way timing cells exchange as described in Chapter 5 but assuming the addition of the quadratic dependence terms, then the clock offset measurement model can be expressed as

$$\begin{aligned} T_{2,r} &= T_{1,r}^2\varphi + T_{1,r}\omega + \phi + d + X_r, \\ T_{3,r} &= T_{4,r}^2\varphi + T_{4,r}\omega + \phi - d - Y_r, \end{aligned}$$

where X_r and Y_r are modeled as exponential RVs, the motivations behind which are explained in detail in Chapter 5. Based on the above model, the likelihood function can be written as

$$\begin{aligned} L\left(d,\phi,\omega,\varphi\right) &= \alpha^{-2N} \\ &\times \exp\left[-\frac{1}{\alpha}\sum_{i=1}^{N}(T_{4,r}^2 - T_{1,r}^2)\varphi + \sum_{i=1}^{N}\{(T_{4,r} - T_{1,r})\omega + (T_{2,r} - T_{3,r}) - 2d\}\right] \\ &\times \prod_{i=1}^{N} I\left[+T_{2,r} - T_{1,r}^2\varphi - T_{1,r}\omega - \phi - d \geq 0\right] \prod_{i=1}^{N} I\left[-T_{3,r} + T_{4,r}^2\varphi + T_{4,r}\omega + \phi - d \geq 0\right]. \end{aligned}$$

We assume that the clocks can neither stop nor run backwards, which implies that the clock skew ω is not less than or equal to 0 and hence is always positive. The actual value of practical clock skew is usually close to 1. Finally, to simplify the derivation, φ has been assumed to be positive. A negative value of φ will result in the same closed form expression of the MLE following a procedure similar to that adopted below.

12.2 THE ESTIMATION PROCEDURE

The constraints present in the likelihood function can be expressed equivalently as

$$\begin{aligned} d &> 0, \quad \varphi > 0, \quad \omega > 0, \\ \infty &> \phi > -\infty, \\ d &\leq +T_{2,i} - T_{1,i}^2\varphi - T_{1,i}\omega - \phi, \quad i = 1, \ldots, N \qquad (12.1) \\ d &\leq -T_{3,j} + T_{4,j}^2\varphi + T_{4,j}\omega + \phi, \quad j = 1, \ldots, N. \qquad (12.2) \end{aligned}$$

These constraints can be viewed as $2N$ four-dimensional curves due to the four unknowns. The three-dimensional region where the two sets of N curves in (12.1) and (12.2) intersect each other yields ϕ in terms of ω and φ as

$$2\phi = (T_{2,i} + T_{3,j}) - (T_{1,i}^2 + T_{4,j}^2)\varphi - (T_{1,i} + T_{4,j})\omega, \quad i, j = 1, \ldots, N. \qquad (12.3)$$

Plugging (12.3) back into (12.1), the sets of constraints can now be expressed as

$$d \leq T_{2,i} - T_{1,i}^2\varphi - T_{1,i}\omega - \frac{1}{2}\left[(T_{2,i} + T_{3,j}) - (T_{1,i}^2 + T_{4,j}^2)\varphi - (T_{1,i} + T_{4,j})\omega\right], \quad i, j = 1, \ldots, N$$

or equivalently,

$$2d \leq (T_{2,i} - T_{3,j}) + (T_{4,j}^2 - T_{1,i}^2)\varphi + (T_{4,j} - T_{1,i})\omega, \quad i, j = 1, \ldots, N. \qquad (12.4)$$

The above inequalities in (12.4) represent a three-dimensional region due to three unknowns consisting of N^2 surfaces forming the boundary of the support region. To find this boundary of the support region as a function of φ only, the intersection of these surfaces in (12.4) with each other is

$$\begin{aligned} \omega &= \frac{\left[(T_{2,k} - T_{3,l}) - (T_{2,i} - T_{3,j})\right] + \left[(T_{4,l}^2 - T_{1,k}^2) - (T_{4,j}^2 - T_{1,i}^2)\right]\varphi}{(T_{4,j} - T_{1,i}) - (T_{4,l} - T_{1,k})} \\ &= u_a + v_a\varphi, \qquad (12.5) \end{aligned}$$

where

$$
\begin{aligned}
u_a &= \frac{(T_{2,k} - T_{3,l}) - (T_{2,i} - T_{3,j})}{(T_{4,j} - T_{1,i}) - (T_{4,l} - T_{1,k})}, \\
v_a &= \frac{(T_{4,l}^2 - T_{1,k}^2) - (T_{4,j}^2 - T_{1,i}^2)}{(T_{4,j} - T_{1,i}) - (T_{4,l} - T_{1,k})},
\end{aligned}
$$

and a is a simplified index notation, being a function of the indices (i, j, k, l). Now plugging (12.5) into (12.4) yields the support region in terms of d as a function of φ only as

$$
\begin{aligned}
2d &\leq (T_{2,m} - T_{3,n}) + (T_{4,n} - T_{1,m}) \frac{(T_{2,p} - T_{3,q}) - (T_{2,m} - T_{3,n})}{(T_{4,n} - T_{1,m}) - (T_{4,q} - T_{1,p})} \\
&+ (T_{4,n}^2 - T_{1,m}^2)\varphi + (T_{4,n} - T_{1,m}) \frac{(T_{4,q}^2 - T_{1,p}^2) - (T_{4,n}^2 - T_{1,m}^2)}{(T_{4,n} - T_{1,m}) - (T_{4,q} - T_{1,p})} \varphi \\
&= \frac{(T_{4,n} - T_{1,m})(T_{2,p} - T_{3,q}) - (T_{2,m} - T_{3,n})(T_{4,q} - T_{1,p})}{(T_{4,n} - T_{1,m}) - (T_{4,q} - T_{1,p})} \\
&+ \frac{(T_{4,n} - T_{1,m})(T_{4,q}^2 - T_{1,p}^2) - (T_{4,n}^2 - T_{1,m}^2)(T_{4,q} - T_{1,p})}{(T_{4,n} - T_{1,m}) - (T_{4,q} - T_{1,p})} \varphi \\
&= w_b + z_b \varphi,
\end{aligned} \tag{12.6}
$$

where

$$
\begin{aligned}
w_b &= \frac{(T_{4,n} - T_{1,m})(T_{2,p} - T_{3,q}) - (T_{2,m} - T_{3,n})(T_{4,q} - T_{1,p})}{(T_{4,n} - T_{1,m}) - (T_{4,q} - T_{1,p})}, \\
z_b &= \frac{(T_{4,n} - T_{1,m})(T_{4,q}^2 - T_{1,p}^2) - (T_{4,n}^2 - T_{1,m}^2)(T_{4,q} - T_{1,p})}{(T_{4,n} - T_{1,m}) - (T_{4,q} - T_{1,p})},
\end{aligned}
$$

and b is again a simplified index notation as a function of the indices (m, n, p, q). Now it is clear that the MLE $(\hat{\varphi}, \hat{d}, \hat{\omega}, \hat{\phi})$ is unique and is given by that intersection of two curves on the boundary of the support region in (12.6) where the term $\sum_{r=1}^{N}\{(T_{4,r}^2 - T_{1,r}^2) + v_a(T_{4,r} - T_{1,r})\} - Nz_b$ is negative for one curve and positive for the other. The MLE $(\hat{\varphi}, \hat{d}, \hat{\omega}, \hat{\phi})$ can be derived by the following observations:

1. It is clear that the MLE lies on the boundary of the support region. This is because for any d lying somewhere within the support region, the likelihood

function (12.1) can be further increased by augmenting d until it reaches the boundary of the support region.

2. Maximizing the likelihood function is equivalent to minimizing the argument of the exponential function $\Phi = \sum_{r=1}^{N}[(T_{4,r}^2 - T_{1,r}^2)\varphi + (T_{4,r} - T_{1,r})\omega + (T_{2,r} - T_{3,r}) - 2d]$ in the likelihood function expression. Therefore, plugging (12.5) and (12.6) into the expression for Φ, it can be written in the form of a set $\phi_{a,b}$ depending on indices a and b as follows:

$$\begin{aligned}\phi_{a,b} &= \sum_{r=1}^{N}\left[(T_{4,r}^2 - T_{1,r}^2)\varphi + (T_{4,r} - T_{1,r})(u_a + v_a\varphi) + (T_{2,r} - T_{3,r}) - (w_b + z_b\varphi)\right], \\ &\propto \left[\sum_{r=1}^{N}\{(T_{4,r}^2 - T_{1,r}^2) + v_a(T_{4,r} - T_{1,r})\} - Nz_b\right]\varphi.\end{aligned}$$

3. Starting from z_b corresponding to $\min_b\{w_b\}$ and evaluating $\phi_{a,b}$ on each subsequent z_b on the boundary of the support region, observe that for each particular segment, $\phi_{a,b}$ can be minimized by taking the largest possible $\hat{\varphi}$ if the term $\sum_{r=1}^{N}\{(T_{4,r}^2 - T_{1,r}^2) + v_a(T_{4,r} - T_{1,r})\} - Nz_b$ is negative and by taking the smallest possible $\hat{\varphi}$ if $\sum_{r=1}^{N}\{(T_{4,r}^2 - T_{1,r}^2) + v_a(T_{4,r} - T_{1,r})\} - Nz_b$ is positive.

4. Since the boundary of the support region is formed by the curves in (12.6) in an order of decreasing slopes $\{z_b\}$, the intersection where the sign of $\sum_{r=1}^{N}\{(T_{4,r}^2 - T_{1,r}^2) + v_a(T_{4,r} - T_{1,r})\} - Nz_b$ (and hence the sign of $\phi_{a,b}$) changes from negative to positive occurs only once. Therefore, the MLE must be unique.

5. Let $c = \min_a\{v_a\}$ and $s = \{a\} \setminus c$. Now comparing $\phi_{c,b}$ and $\phi_{s,b}$ on the boundary of the support region yields the following three options:

   - The signs of both $\phi_{s,b}$ and $\phi_{c,b}$ change at the same intersection of curves in (12.6). In this case, $\phi_{c,b} < \phi_{s,b}$ since $v_c < v_s$.
   - The sign change for $\phi_{s,b}$ occurs at an intersection of the curves in (12.6) to the right of the intersection where the sign change for $\phi_{c,b}$ occurs. This is not possible because for the same z_b, $\phi_{s,b}$ must have a sign change at or to the left of the intersection where the same occurs for $\phi_{c,b}$.
   - The sign of $\phi_{s,b}$ changes at an intersection of curves in (12.6) which is to the left of the intersection where the sign change for $\phi_{c,b}$ occurs. Now even on the first intersection, $\phi_{c,b} < \phi_{s,b}$ since $v_c < v_s$. Due to the continuity of

$\phi_{c,b}$ (and hence the continuity of the likelihood function) on the support region, $\phi_{c,b}$ can be further decreased by increasing φ until it touches the second intersection. Therefore, $a = c$ should be used to find the index b corresponding to the minimum of the set $\phi_{c,b}$.

6. Finally, in the light of above observations, by checking the sign of the expression $\sum_{r=1}^{N} \{(T_{4,r}^2 - T_{1,r}^2) + v_c(T_{4,r} - T_{1,r})\} - Nz_b$ for each b, we conclude that the MLE $\hat{\varphi}$ can be expressed as

$$\hat{\varphi} = \left[\frac{D_{(4,n)(1,m)}D_{(2,p)(3,q)} - D_{(2,m)(3,n)}D_{(4,q)(1,p)}}{D_{(4,n)(1,m)} - D_{(4,q)(1,p)}} - \frac{D_{(4,j)(1,i)}D_{(2,k)(3,l)} - D_{(2,i)(3,j)}D_{(4,l)(1,k)}}{D_{(4,j)(1,i)} - D_{(4,l)(1,k)}}\right] \Bigg/ \left[\frac{D_{(4,j)(1,i)}D^2_{(4,l)(1,k)} - D^2_{(4,j)(1,i)}D_{(4,l)(1,k)}}{D_{(4,j)(1,i)} - D_{(4,l)(1,k)}} - \frac{D_{(4,n)(1,m)}D^2_{(4,q)(1,p)} - D^2_{(4,n)(1,m)}D_{(4,q)(1,p)}}{D_{(4,n)(1,m)} - D_{(4,q)(1,p)}}\right], \tag{12.7}$$

where the difference between any two timestamps is denoted by their indices as $D_{(,)(,)}$ (e.g., $T_{4,n} - T_{1,m}$ as $D_{(4,n)(1,m)}$) and the square of timestamps as $D^2_{(,)(,)}$ (e.g., $T_{4,j}^2 - T_{1,i}^2$ as $D^2_{(4,j)(1,i)}$), where the indices $\{i, j, k, l, m, n, p, q\}$ correspond to the two sets of curves in (12.6) for which the sign of $\sum_{r=1}^{N} \{(T_{4,r}^2 - T_{1,r}^2) + v_c(T_{4,r} - T_{1,r})\} - Nz_b$ changes from negative to positive. Consequently, plugging $\hat{\varphi}$ in (12.6), (12.5), and (12.3), we can express $\hat{d}, \hat{\omega}$, and $\hat{\phi}$ as follows:

$$\begin{aligned} \hat{d} &= \frac{1}{2}\frac{D_{(4,j)(1,i)}D_{(2,k)(3,l)} - D_{(2,i)(3,j)}D_{(4,l)(1,k)}}{D_{(4,j)(1,i)} - D_{(4,l)(1,k)}} \\ &\quad + \frac{1}{2}\frac{D_{(4,j)(1,i)}D^2_{(4,l)(1,k)} - D^2_{(4,j)(1,i)}D_{(4,l)(1,k)}}{D_{(4,j)(1,i)} - D_{(4,l)(1,k)}}\hat{\varphi}, \\ \hat{\omega} &= \frac{D_{(2,k)(3,l)} - D_{(2,i)(3,j)}}{D_{(4,j)(1,i)} - D_{(4,l)(1,k)}} + \frac{\left[D^2_{(4,l)(1,k)} - D^2_{(4,j)(1,i)}\right]}{D_{(4,j)(1,i)} - D_{(4,l)(1,k)}}\hat{\varphi}, \\ \hat{\phi} &= \frac{T_{2,i} + T_{3,j} - (T_{1,i}^2 + T_{4,j}^2)\hat{\varphi} - (T_{1,i} + T_{4,j})\hat{\omega}}{2}. \end{aligned}$$

Algorithm 6 presents in detail the steps required to find this MLE ($\hat{\varphi}$, $\hat{d}, \hat{\omega}$, $\hat{\phi}$). As N becomes large, clock drift estimation becomes particularly useful for capturing the clock dynamics in a better way. The complete procedure for finding this

Algorithm 6: MLE for $\hat{\varphi}$, d, $\hat{\omega}$, and $\hat{\phi}$

Compute the set $\{v_a\}$ and $\{z_b\}$;

$c = \min_a\{v_a\}$;

$(i,j,k,l) \longrightarrow \min\{w_b\}$;

LABEL:

$$\varphi^{m,n,p,q} = \left[\frac{D_{(4,n)(1,m)}D_{(2,p)(3,q)} - D_{(2,m)(3,n)}D_{(4,q)(1,p)}}{D_{(4,n)(1,m)} - D_{(4,q)(1,p)}} - \frac{D_{(4,j)(1,i)}D_{(2,k)(3,l)} - D_{(2,i)(3,j)}D_{(4,l)(1,k)}}{D_{(4,j)(1,i)} - D_{(4,l)(1,k)}}\right] \Big/ \left[\frac{D_{(4,j)(1,i)}D^2_{(4,l)(1,k)} - D^2_{(4,j)(1,i)}D_{(4,l)(1,k)}}{D_{(4,j)(1,i)} - D_{(4,l)(1,k)}} - \frac{D_{(4,n)(1,m)}D^2_{(4,q)(1,p)} - D^2_{(4,n)(1,m)}D_{(4,q)(1,p)}}{D_{(4,n)(1,m)} - D_{(4,q)(1,p)}}\right];$$

$(e,f,g,h) = \arg\min_{m,n,p,q}\{\varphi^{m,n,p,q}\}$;

if $\left[\sum_{r=1}^{N}\{D^2_{(4,r)(1,r)} + v_c D_{(4,r)(1,r)}\} - Nz_b\right]^{i,j,k,l} \times \left[\sum_{r=1}^{N}\{D^2_{(4,r)(1,r)} + v_c D_{(4,r)(1,r)}\} - Nz_b\right]^{e,f,g,h} < 0$ **then**

$$\hat{\varphi} = \varphi^{e,f,g,h};$$

$$\hat{d} = \frac{1}{2}\frac{D_{(4,f)(1,e)}D_{(2,g)(3,h)} - D_{(2,e)(3,f)}D_{(4,h)(1,g)}}{D_{(4,f)(1,e)} - D_{(4,h)(1,g)}} + \frac{1}{2}\frac{D_{(4,f)(1,e)}D^2_{(4,h)(1,g)} - D^2_{(4,f)(1,e)}D_{(4,h)(1,g)}}{D_{(4,f)(1,e)} - D_{(4,h)(1,g)}}\hat{\varphi},$$

$$\hat{\omega} = \frac{D_{(2,g)(3,h)} - D_{(2,e)(3,f)} + \left[D^2_{(4,h)(1,g)} - D^2_{(4,f)(1,e)}\right]}{D_{(4,f)(1,e)} - D_{(4,h)(1,g)}},$$

$$\hat{\phi} = \frac{1}{2}\left[T_{2,e} + T_{3,f} - (T^2_{1,e} + T^2_{4,f})\hat{\varphi}\right] - \frac{1}{2}\left[(T_{1,e} + T_{4,f})\hat{\omega}\right];$$

else

Discard (i,j,k,l) curve;

$(i,j,k,l) = (e,f,g,h)$;

goto LABEL;

end if

MLE $(\hat{\varphi}, \hat{d}, \hat{\omega}, \hat{\phi})$ is explained in Algorithm 6. Algorithm 6 starts from the curve in (12.6) for which w has the least value. It selects the intersection of this curve with the neighboring curve intersecting it, and it checks the sign change condition of $\sum_{r=1}^{N}\{(T^2_{4,r} - T^2_{1,r}) + v_c(T_{4,r} - T_{1,r})\} - Nz_b$. If the condition is not satisfied, the first

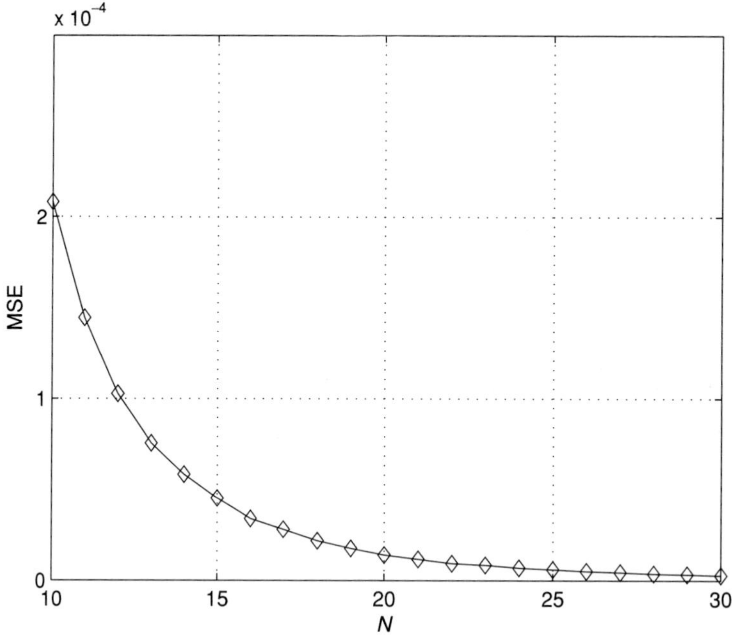

Figure 12.1: MSE of $\hat{\varphi}$ as a function of number of observations N.

curve is discarded and the same procedure is repeated for the second curve and so on until the same condition is satisfied.

Figure 12.1 shows the MSE of $\hat{\theta}_D$ as a function of the number of timing messages N. It is evident that the MLE performs well even for small N and hence is suitable for the power-limited regime of WSNs.

Chapter 13

JOINT SYNCHRONIZATION OF CLOCK OFFSET AND SKEW IN A RECEIVER–RECEIVER PROTOCOL

Turning our attention in this chapter to a general receiver–receiver protocol, we address the synchronization problem in which a master node sends reference broadcasts to the neighboring nodes (e.g., RBS [22]). As discussed earlier, the main advantage of adopting such an approach is that *all* the deterministic and non-deterministic delay components on the sender side (such as send time, transmission time, channel access time) are eliminated and hence the clocks of the beacon-receiving nodes can be very tightly synchronized. The importance of RRS increases due to the fact that the channel access time at the MAC layer is the largest source of error in solving a synchronization problem. This chapter applies both classical and Bayesian estimation approaches to synchronize a set of nodes receiving timing messages from a master node.

The main topics in this chapter are as follows. First, the JMLE for clock phase offset and skew under the exponential noise model is formulated and found via a direct algorithm. Second, the Gibbs sampler is proposed for joint clock phase offset and skew estimation and is shown to provide superior performance relative to JMLE. Finally, lower and upper bounds for the MSE of JMLE and Gibbs sampler are introduced in terms of the MSE of the MVUE and the conventional BLUE, respectively.

13.1 Modeling Assumptions

In a general receiver–receiver protocol, a transmitter node broadcasts N synchronization signals and the receiver nodes put timestamps on these signals. Then, for efficient implementation, the receivers pass the data consisting of the timestamps to the transmitter where the clock offsets and skews between different pairs of nodes are calculated. With the help of this protocol, two of the main error sources of clock synchronization are eliminated, which are uncertainties in *send time* and *access time*. Furthermore, the difference between propagation times is negligible compared to the uncertainty in *receive time*, which becomes the only error source. Therefore, the ith timestamps at the receivers X and Y are given by

$$X[i] = T_1 + \phi_x + \omega_x \tau[i] + v_{x,\lambda_x}[i], \quad Y[i] = T_1 + \phi_y + \omega_y \tau[i] + v_{y,\lambda_y}[i], \tag{13.1}$$

where T_1 stands for the time on the transmitter when it sends the first synchronization signal, ϕ_x and ω_x denote the offset and skew between the clocks of the receiver X and the transmitter, $\tau[i]$ is the difference between T_1 and the time of ith synchronization signal (with respect to the transmitter's clock), and $v_{x,\lambda_x}[i]$ is the exponential iid noise (with mean $1/\lambda_x$), with $i = 1, \ldots, N$. The parameters to be estimated, the offset and skew between the clocks of the nodes X and Y, are given by the following equations:

$$\Phi = \phi_x - \phi_y \,, \quad \omega = \omega_x - \omega_y \,. \tag{13.2}$$

13.2 JOINT MAXIMUM LIKELIHOOD ESTIMATION (JMLE) OF THE OFFSET AND SKEW

The estimation of clock skew is more important in the context of energy-constrained sensor networks. In [83] it is shown that under uniform noise, there are infinite solutions for MLE. Moreover, the support of the likelihood function is not convex, which leaves out the possibility of taking the mean of all equally likely solutions. In this section, we will consider the modeling framework described in (13.1). As long as the two parameter sets $\{\phi_x, \omega_x, \lambda_x\}$ and $\{\phi_y, \omega_y, \lambda_y\}$ do not have a direct relationship and the noise sources in different nodes are independent (both of which are realistic assumptions), we can find the JML estimator for Φ and ω without any loss of information by estimating the parameters (ϕ_x, ω_x) and (ϕ_y, ω_y) separately and plugging these estimates back into (13.2). Thus, we will concentrate on the estimation of ϕ_x and ω_x. First, for simplicity, we will assume that $\tau[i] = i - 1$ and $T_1 = 0$, then the likelihood function becomes

$$\begin{aligned} L(\phi_x, \omega_x) &= \prod_{i=1}^{N} \lambda_x \exp\{-\lambda_x (X[i] - (\phi_x + (i-1)\omega_x))\} \, I_{(X[i] \geq \phi_x + (i-1)\omega_x)} \\ &= \lambda_x^N \exp\left[-\lambda_x N(\overline{X} - f)\right] \prod_{i=1}^{N} I_{(X[i] \geq f_i)} \,, \end{aligned} \tag{13.3}$$

where $f(\phi_x, \omega_x) = \phi_x + \frac{N-1}{2}\omega_x$, $f_i(\phi_x, \omega_x) = \phi_x + (i-1)\omega_x$, $\overline{X}$ stands for the sample

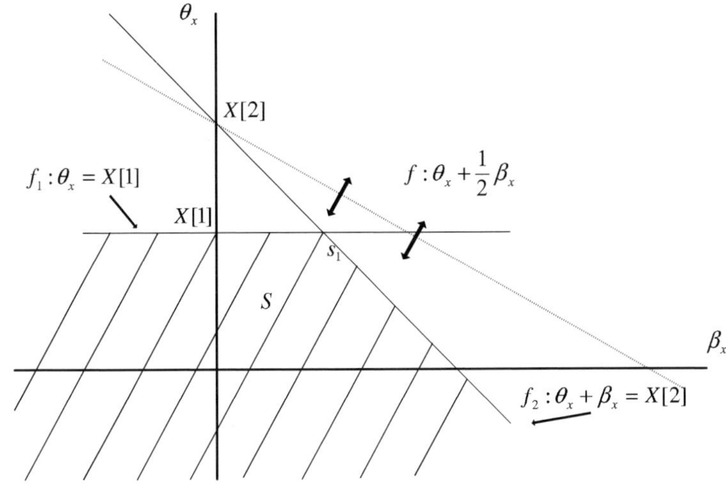

Figure 13.1: S and the solution s_1.

mean of observations $X[i]$ ($i = 1, \ldots, N$), and $I_{(x \geq a)}$ denotes the indicator function, being equal to 1 when $x \geq a$ and 0 elsewhere.

Note that in (13.3), the multiplication of indicator functions defines a convex region ($\mathbf{S}$) on the parameter space (ϕ_x, ω_x), with $\mathbf{S} = \{(\phi_x, \omega_x) : \bigcap_{i=1}^{N} f_i(\phi_x, \omega_x) \leq X[i]\}$ $\mathbf{S}$ has k vertices $\{s_j\}_{j=1}^{k}$ and $k+1$ edges ($1 \leq k \leq N-1$). Specifically, the shape of this region and the value of k strongly depend on the ordering of $X[1], \ldots, X[N]$. On this region, we have to maximize the objective function, $f(\phi_x, \omega_x) = \phi_x + \frac{N-1}{2}\omega_x$. Since $0 < \frac{N-1}{2} < N-1$, the support of the solution is guaranteed to be a closed-convex region on the boundary of $\mathbf{S}$. If $N = 2m$, the solution will be one of the vertices s_j and if $N = 2m-1$ the solution will assume possibly a segment of the line $f_m : \phi_x + (m-1)\omega_x$ (or again one of the vertices s_j, depending on the ordering of the observations). Figure 13.1 illustrates these remarks for $N = 2$ and $X[2] > X[1]$. In this illustrative example, since f attains its maximum on s_1 amongst all points on $\mathbf{S}$, s_1 gives the JMLE of ϕ_x and ω_x. Before proceeding any further, we have to clarify one more point. In derivations up to now, we assumed that λ_x and λ_y were both known. However, if we assume they are unknown and use the reduced likelihood function for (ϕ_x, ω_x) as in [41], it is straightforward to show that we end up with the same JML solution. Therefore, knowledge of λ_x and λ_y is not required to find the MLE.

13.3 Application of the Gibbs Sampler

Although it is possible to find the exact solution for the ML estimate as explained above, we will also apply the Gibbs sampler to jointly estimate the parameters [54],

[78]. Although by using the Gibbs sampler it is possible to find an approximate JMLE which is arbitrarily close to the exact one, there are some more important advantages that the Gibbs sampler provides. First, it can be shown that the JMLE $(\hat{\phi}_{x,ML}, \hat{\omega}_{x,ML})$ is biased for finite N. (As an example consider the case in Figure 13.1, $E[\hat{\phi}_{x,ML}] = E[X[1]] = E[\phi_x + v_{x,\lambda_x}[1]] = \phi_x + 1/\lambda_x$.) For this reason, we need to look for an MVU estimator. However, the Neyman–Fisher factorization theorem provides $\min_i((X[i] - \phi_x + (i-1)\omega_x))$ as sufficient statistics, which is not independent of the parameters to be estimated. On the other hand, if we use the Gibbs sampler at the end we do not have just a single point estimate but the posterior distribution for the parameters to be estimated as the output. Then, we can either find the JML estimator or set the corresponding estimator as the mean value of the posterior distribution of the parameter, which will automatically perform the marginalization and will give better results with reduced bias and variance. Another appealing feature of the Gibbs sampler is its straightforward extendability for additional unknown parameters. For example, it is possible that λ_x is unknown or in addition to the clock phase offset and skew we could have clock drifts: γ_x and γ_y. The drifts will be observed on the right hand side of (13.1) as additional terms: $\tau^2[i]\gamma_x$ and $\tau^2[i]\gamma_y$. It is straightforward to adapt the Gibbs sampler to these scenarios. However, here we limit the application of Gibbs sampler to the estimation problem depicted by (13.1).

Before applying the Gibbs sampler, we will briefly review some information about Gibbs sampling. Assume that we have the data vector $\mathbf{z}$ and we want to estimate some parameters $\boldsymbol{\Theta} = [\theta_1, \theta_2, \ldots, \theta_M]^T$. For any kind of statistical inference we want to use the joint posterior distribution of the parameters $p(\Theta|\mathbf{z}) \propto p(\mathbf{z}|\Theta)p(\Theta)$ (in point estimation, the prior distribution $p(\Theta)$ is chosen as non-informative). When it is hard to carry out mathematical derivations on the posterior, we stick to Monte-Carlo methods, i.e., draw as many samples as possible from the posterior so that the inference we make using these samples is arbitrarily close to the exact solution. When it is hard to draw samples from the joint posterior directly, Markov Chain Monte-Carlo (MCMC) type iterative methods are used [54], [78]. This resumes to setting up a Markov chain whose stationary distribution is the joint posterior we need. One convenient way to do this is to use Gibbs sampling in which we iteratively draw samples from one-dimensional conditionals $p(\theta_i|\mathbf{z}, \overline{\boldsymbol{\Theta}_\mathrm{i}})$, where $\overline{\boldsymbol{\Theta}_\mathrm{i}}$ is an $(M-1) \times 1$ vector with entries $\{\theta_j\}_{j \neq i}$. Under mild conditions, these one-dimensional conditional distributions uniquely determine the joint posterior distribution [10].

Specifically, the general algorithm for Gibbs sampling with initial values $\Theta^{(0)} = [\theta_1^{(0)}, \ldots, \theta_M^{(0)}]$ is to iterate the following steps:

- Draw $\theta_1^{(1)}$ from $p(\theta_1|\mathbf{z}, \theta_2^{(0)}, \ldots, \theta_M^{(0)})$.
- Draw $\theta_2^{(1)}$ from $p(\theta_2|\mathbf{z}, \theta_1^{(1)}, \theta_3^{(0)}, \ldots, \theta_M^{(0)})$.

$\vdots$

- Draw $\theta_M^{(1)}$ from $p(\theta_M|\mathbf{z}, \theta_1^{(1)}, \ldots, \theta_{M-1}^{(1)})$.

After a threshold value t, the set $\{\Theta^{(t)}, \Theta^{(t+1)}, \ldots\}$ behaves like samples from the joint posterior of the parameters.

One important point is that the joint posterior distribution should be proper. Otherwise the Gibbs sampler always converges to some local points, but not necessarily to a meaningful one [35]. For this reason, to ensure that the posterior is proper, in the application of Gibbs sampler to the point estimation, priors are not directly chosen as flat, but they are chosen from conjugate families and then their parameters arranged so as to have non-informative priors. However, in our case, the likelihood function itself can be used as the posterior distribution, since its integral is always bounded and positive-valued which makes it proper. We do not need to use any other type of priors but flat. Then in our case, using (13.3), the procedure becomes

- Draw $\phi_x^{(1)}$ from $\propto \exp(\lambda_x N \phi_x) I(\phi_x \leq \min_i(X[i] - (i-1)\omega_x^{(0)}))$.
- Draw $\omega_x^{(1)}$ from $\propto \exp\left[\lambda_x \frac{N(N-1)}{2} \omega_x\right] I(\omega_x \leq \min_i(\frac{X[i]-\phi_x^{(1)}}{i-1}))$.

For ϕ_x^{t+1}, we will draw a sample from the exponential distribution with parameter $\lambda_x N$, multiply it by -1 and add $\min_i(X[i] - (i-1)\omega_x^{(t)})$ to it. The procedure for ω_x^{t+1} is similar. Note that if λ_x were unknown, we would utilize the Gamma distribution to draw for $\lambda_x^{(t+1)}$.

13.4 PERFORMANCE BOUNDS AND SIMULATIONS

In this section, we will look at the performances of the Gibbs sampler and the JML estimator. However, it will be useful to have some benchmarks with which to compare their performances. First, we will look for lower bounds. Since the likelihood function does not satisfy the regularity conditions required by the CRLB, calculating CRLBs is not considered. One possible lower bound can be found by assuming that all the parameters are known but the one to be bounded, which reduces the problem to the well-known derivation of the bound for a single unknown parameter in exponential noise. Then we can find the MVUE both for the phase offset and skew in closed forms. For ϕ_x, the MVUE then becomes

$$\hat{\phi}_{x,MVUE} = \min_i(X[i] - (i-1)\omega_x) - \frac{1}{N\lambda_x}, \tag{13.4}$$

and the MSE of the estimator equals $1/(N\lambda_x)^2$. For ω_x, the likelihood function is

$$L(\omega_x) = C\exp\left[\lambda_x \frac{N(N-1)}{2}\omega_x\right]\prod_{i=2}^{N} I\left(\omega_x \le \frac{X[i]-\phi_x}{i-1}\right). \tag{13.5}$$

From the factorization theorem [44], $\min_i(\frac{X[i]-\phi_x}{i-1})$ is sufficient statistics and it is straightforward to show that it is also complete. This result can be established by following a similar proof to that in [50] for ϕ_x. Then, by the Lehmann–Scheffé theorem, the MVUE for the skew when the offset and λ_x are known takes the form

$$\hat{\omega}_{x,MVUE} = \min_i\left(\frac{X[i]-\phi_x}{i-1}\right) - \frac{2}{\lambda_x N(N-1)}. \tag{13.6}$$

The MSE of the estimator (13.6) is equal to the variance of $Z = \min_{i=2,3,\ldots,N}(\frac{X[i]-\phi_x}{i-1})$. Thus, we first need to determine the distribution of Z. From the theory of order statistics, the distribution of the minimum of a sample set is given by $F(z) = 1 - (1-F_2(z))(1-F_3(z))\ldots(1-F_N(z))$, where $F_i(z) = Pr(\frac{X[i]-\phi_x}{i-1} \le z) = Pr(v_{x,\lambda_x}[i] \le (i-1)(z-\omega_x)) = (1-\exp[\lambda_x(i-1)(z-\omega_x)])I(z \ge \omega_x)$. Then the distribution becomes

$$F(z) = 1 - \exp[\lambda_x(zi\omega_x)(1+2+\cdots+N-1)] = 1 - \exp\left[\lambda_x(z-\omega_x)\frac{N(N-1)}{2}\right], \tag{13.7}$$

which is an exponential distribution with the scale parameter $\frac{\lambda_x N(N-1)}{2}$ and the location parameter ω_x. The MSE of $\hat{\omega}_{x,MVUE}$ equals the variance of Z which is $4/(\lambda_x N(N-1))^2$. Therefore, we do not expect the MSE of the joint estimator for (ϕ_x, ω_x) to decay faster than $O(1/N^2, 1/N^4)$.

We will also consider the BLUE, since it will represent an upper bound. Here, the same notation is used as [83] except that $\mathbf{X}$ is replaced by $\mathbf{A}$ ($\mathbf{A} \triangleq [\mathbf{1}, \mathbf{x}]$, where $\mathbf{1} = [1,1,\cdots,1]^T$ and $\mathbf{x} = [0,1,\cdots,N-1]^T$) to prevent possible confusion. Since noise is not zero mean in our model, we need to subtract $1/\lambda_x$ from the resulting

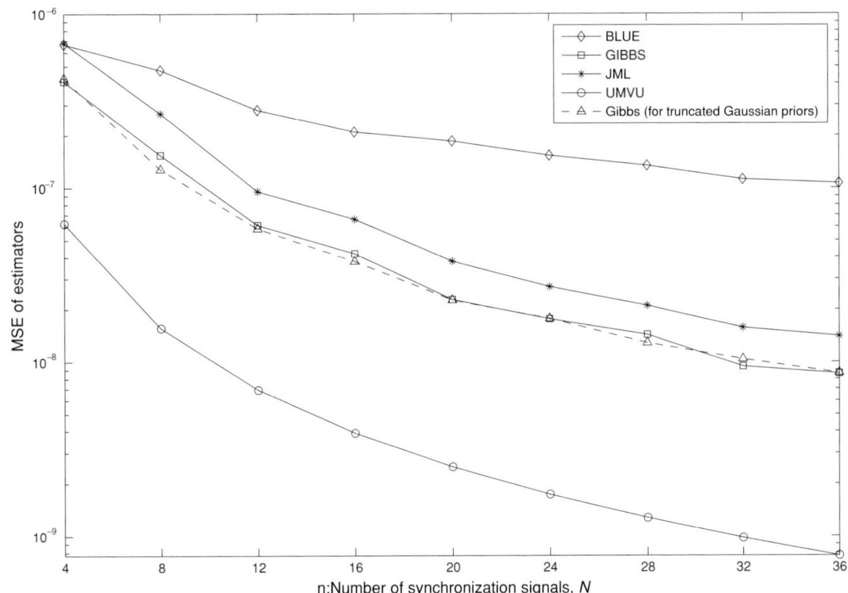

Figure 13.2: MSE for $\hat{\phi}_{x,BLUE}$, $\hat{\phi}_{x,JML}$, $\hat{\phi}_{x,Gibbs}$, and $\hat{\phi}_{x,MVUE}$.

linear estimate of ϕ_x so as to end up with the BLUE. Then we have

$$[\hat{\phi}_{x,BLUE},\ \hat{\omega}_{x,BLUE}]^T = (\mathbf{A}^T\mathbf{A})^{-1}\mathbf{A}^T\mathbf{X} - \left[\frac{1}{\lambda_x},\ 0\right]^T.$$

It is known that $\text{var}([\hat{\phi}_{x,BLUE},\ \hat{\omega}_{x,BLUE}]^T) = \text{diag}\{1/\lambda_x^2(\mathbf{A}^T\mathbf{A})^{-1}\} \propto [1/N,\ 1/N^3]^T$. The MSE of the Gibbs sampler and the JML estimator for $\phi_x = 1$ and $\omega_x = 0.01$ with $\lambda_x = 10^3$ (which makes $\text{var}(v_{x,\lambda_x}) = 10^{-6}$) are presented in Figures 13.2 and 13.3, respectively. In these simulations, the initial values of clock parameters are chosen as zeros. These figures also include the lower and upper bounds presented above. The MSEs are plotted against the number of synchronization signals from 4 to 36. It is interesting to note that the MSE of the Gibbs sampler and the JML estimator behave like the lower bound, i.e., decay rates on the order of $O(1/N^2)$ and $O(1/N^4)$, respectively. Note also that the Gibbs sampler performs better with MSE values reduced around 40% for ϕ_x and 25% for ω_x compared to the corresponding values for the JML estimator. We should also note that the convergence of the Gibbs sampler is achieved after a number of iterations on the order of N. To shed some light on the sensitivity of the Gibbs sampler to the prior mismatch, we have also provided some simulation results for the mismatched prior knowledge. This is important for engineers and system designers in order to make a proper choice of estimator for their systems. Figures 13.2 and 13.3 show the performance of the Gibbs sampler where we have modeled the actual prior as a truncated Gaussian while the assumed prior

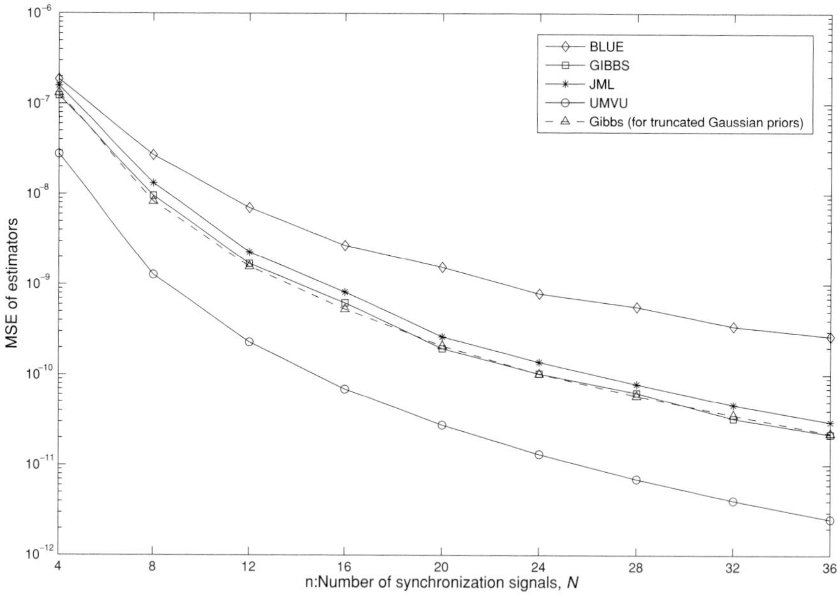

Figure 13.3: MSE for $\hat{\omega}_{x,BLUE}$, $\hat{\omega}_{x,JML}$, $\hat{\omega}_{x,Gibbs}$, and $\hat{\omega}_{x,MVUE}$.

in the Gibbs sampler is uniform. For the prior of the offset, truncation points have been chosen as 0 and 10 whereas the mean and standard deviation of parent Gaussian distribution are 5 and 2, respectively. And for the prior of the skew, truncation points are chosen as 0 and 1, whereas the mean and standard deviation of parent Gaussian distribution are 0.5 and 0.25, respectively.

One drawback of the Gibbs sampler is definitely its computational complexity. The computational complexity of the Gibbs sampler is affected by the random number generations in each iteration and the number of iterations necessary to converge. Although the Gibbs sampler clearly requires more computations than other methods, the required level of precision can be achieved by fewer signal transmissions. Hence, there is a tradeoff between complexity and the performance achieved by the Gibbs sampler.

Chapter 14

ROBUST ESTIMATION OF CLOCK OFFSET

Clock synchronization between any two nodes is generally accomplished through message exchanges. Due to the presence of non-deterministic and possibly unbounded message delays, messages can be delayed arbitrarily, which makes the clock synchronization very difficult [2]. The most commonly proposed non-deterministic network delay distributions are the Gaussian, exponential, Gamma, and Wei-bull pdfs, see e.g., [1], [52], [70]. In general, it is difficult, if not impossible, to assess which distribution model may be fit to capture the network delay distributions in a given WSN. This is due to the fact that various factors may impact the distribution of network delays differently [33], [49], [92]. The Gaussian pdf [22] and the exponential pdf [1] have also been proposed to model the network delays in WSNs. Here, the ML estimators for clock offset estimation in the presence of Gaussian and exponential network delay distributions will be referred to as the Gaussian ML (GML) and exponential ML (EML), respectively. The simulation results in Figure 6.4 showed that GML and EML are quite sensitive to the network delay distributions. Therefore, one important problem to cope with is the design of clock offset estimation schemes that are robust with respect to the distribution of the unknown network delays.

This chapter deals with the development of clock offset estimators for WSNs that are robust with respect to the possible asymmetries and the unknown or possibly time-varying distributions of the network delays in the uplink and downlink of message exchanges. The two-way message exchange mechanism used in the NTP [61] and TPSN [29] is adopted here as the clock synchronization approach between two nodes of the WSN. To overcome the challenges raised by the unknown distribution of network delays, in this chapter a family of novel clock offset estimation methods, referred to as composite particle filters (CPFs), is proposed. Two general clock offset estimation methods will be designed and their performances thoroughly tested via computer simulations, namely the Gaussian mixture Kalman particle filter (GMKPF) and the CPF without and with bootstrap sampling (BS).

GMKPF combines the importance sampling (IS) based measurement update step with a Kalman filter (KF) based Gaussian sum filter for the time update and proposal density generation. Since GMKPF employs new observations and exploits the expectation–maximization (EM) algorithm to obtain the Gaussian mixture model (GMM), GMKPF is expected to exhibit better estimation performance when compared to GML and EML in general non-Gaussian and non-exponential delay models, respectively. It will be shown via computer simulations that GMKPF yields much more accurate results relative to GML (EML) when the network delays are modeled in terms of a single non-Gaussian (non-exponential) distribution or as a mixture of several distributions. Therefore, the proposed GMKPF method is a high-performance and very reliable clock offset estimation scheme fit to overcome the uncertainties caused by the network delay distributions.

CPF approximates the filtering and predictive distributions by using weighted Gaussian mixtures and is basically implemented via banks of KFs instead of Gaussian particle filters (GPFs) [48]. CPF is a variation of the Gaussian sum particle filter (GSPF) [48], and is fit for the estimation of linear models perturbed by non-Gaussian random noise components. In addition, a variant of the CPF approach based on BS is shown to exhibit good performance in the presence of a small number of observations. The idea behind the CPF with BS is to generate sampled observation data from the original observation data by using the BS, and then to estimate the clock offset using the CPF. Computer simulations illustrate that when there is a reduced number of observations, the CPF with BS presents the best performance when compared to the basic CPF, GML, and EML estimators under general random network delay models such as asymmetric Gaussian, exponential, Gamma, Weibull as well as mixture distributions. The simulation results also corroborate the superior performance of CPF relative to GML and EML for clock offset estimation in general non-Gaussian random delays.

As explained in Chapters 1 and 2, energy conservation is a very important concern. Pottie and Kaiser [75] pointed out that much less power is consumed in processing data than in transmitting it. In fact, they showed that the energy required for a sensor node to transmit 1 bit over 100 meters (3 joules) in a particular WSN is equivalent to the energy required to execute 3 million instructions. Therefore, trading off computational power for more savings in energy consumption appears to be a very feasible approach. Thus, one way to reduce the amount of energy spent on signal transmissions and implicitly on achieving clock synchronization is the use of possibly more sophisticated signal processing algorithms with the goal of achieving more accurate clock offset estimates especially in operational regimes characterized by a reduced number of observations and unknown network delay distributions. When message exchange errors occur, a node will not retransmit the message to a neighbor node but will resample the observation data from the original observation data using the BS approach [21], [101]. The BS can be implemented by constructing a number of resamples of the observed data, each of which is obtained by random sampling with a replacement from the original observed data. Notice that a node will then estimate the clock offset from the resampled observation data by using the CPF. The computer simulations highlight that the CPF with BS exhibits very good performance in various random delay models and it is aimed at reducing the number of message exchanges. Therefore, the CPF with BS leads to less power consumption relative to the CPF, GML, and EML, and is a practically feasible estimator.

It is interesting also to note that there is very little in the literature for clock synchronization in WSNs that addresses the robustness or improving the MSE performance of existing state-of-the-art clock synchronization algorithms in the presence of message errors, unknown and possibly time-varying network delay distributions, or a reduced number of observations (data measurements). Thus far, it appears that only very few preliminary and straightforward applications of standard Kalman filtering or general adaptive signal processing techniques have been reported (see [28], [30], and [95]) to improve the MSE performance of protocols such as RBS [22] or TPSN [29]. However, no attempts have been made to address the problem of building clock synchronization algorithms that are robust to the unknown distribution of random network delays, message errors, or fewer observations. It is the aim in this chapter to answer these points from the light of a composite particle filtering framework.

The main topics in this chapter are as follows. First, a general state-space model is introduced and is used throughout this chapter to formulate the problem and to derive the proposed sequential Bayesian estimation algorithms. Next, a description of the GMKPF approach for estimating the clock offset in WSNs and illustrative computer simulation results are presented to assess the performance of the GMKPF-based clock offset estimator. The CPF and BS-based CPF approaches for estimating the clock offset in WSNs are also developed. In addition, computer simulation results corroborating the superior performance of CPF and BS-based CPF approaches in various network delay distributions are presented.

14.1 PROBLEM MODELING AND OBJECTIVES

Here, we will adopt the two-way timing message exchange protocol depicted in Figure 5.1 as the signaling mechanism for achieving the clock synchronization between two generic nodes, A and B, of a WSN. Under this protocol, the synchronization between two generic nodes A and B is achieved by transmitting timing messages in both directions. The message exchanges between nodes A and B are organized in terms of cycles, and during each cycle a message exchange occurs in each direction. For example, during the kth cycle, node A sends its time reading $T_{1,k}$ to node B, which records the time of arrival of the message sent by node A as $T_{2,k}$, according to its own timescale. Similarly, a timing message exchange is performed from node B to node A. At time $T_{3,k}$ node B transmits back to node A the timing information $T_{2,k}$ and $T_{3,k}$. According to node A's clock, the message transmitted by node B arrives at node A at time $T_{4,k}$. Therefore, at the end of the kth cycle, node A has access to all the time information $\{T_{j,k}\}$, $j = 1, \ldots, 4$, which proves to be sufficient for estimating the clock offset and deterministic propagation delay.

Using the same notation as in Chapter 5, the time differences corresponding to the kth uplink and downlink delay observations of the kth timing message exchange can be expressed as $U_k = T_{2,k} - T_{1,k} = d + \phi + X_k$ and $V_k = T_{4,k} - T_{3,k} = d - \phi + Y_k$, respectively. The fixed value ϕ denotes the clock offset between the two nodes, d stands for the (deterministic) propagation delay, and X_k and Y_k model the variable portions of delay, and might assume any distribution such as Gaussian, exponential, Gamma, Weibull or a mixture of two distributions. For notational convenience, we adopt the notation $x_k = \phi$ for the unknown clock offset. Given the observation samples $\mathbf{z}_k = [U_k, V_k]^T$, our goal is to find the minimum variance estimator of the unknown clock offset ϕ, which is given by

$$\hat{x}_k = E\{x_k | \mathbf{Z}^l\} , \tag{14.1}$$

where $\mathbf{Z}^l$ denotes the set of observed samples up to time l, $\mathbf{Z}^l = \{\mathbf{z}_0, \mathbf{z}_1, \ldots, \mathbf{z}_l\}$. Since the clock offset value is constant, the clock offset is assumed to obey a Gauss–Markov dynamic channel model of the form:

$$x_{k+1} = F x_k + v_k , \tag{14.2}$$

where F represents the state transition matrix for the clock offset. The noise vector v_k is modeled as a Gaussian random variable with zero mean and covariance $E v_k v_k^T = Q$. Notice also that the vector observation model follows from the observed data and can be expressed as

$$\mathbf{z}_{k+1} = \begin{bmatrix} d + x_k + X_k \\ d - x_k + Y_k \end{bmatrix} = \begin{bmatrix} 1 \\ 1 \end{bmatrix} d + \begin{bmatrix} 1 \\ -1 \end{bmatrix} x_k + \mathbf{n}_k , \tag{14.3}$$

where the observation noise vector $\mathbf{n}_k = [X_k, Y_k]^T$ may assume any pdf. It follows from the above discussion that our initial problem is now cast as the estimation of a Gauss–Markov model with unknown states (see (14.2) and (14.3)).

14.2 Gaussian Mixture Kalman Particle Filter (GMKPF)

Particle filtering is a sequential Monte-Carlo sampling method built within the Bayesian paradigm. From a Bayesian perspective, at time k, the posterior distribution $p(x_k|\mathbf{z}_{0:k})$ is the main entity of interest. However, due to the non-Gaussianity of the model (14.3), the analytical expression of $p(x_k|\mathbf{z}_{0:k})$ cannot be obtained as a closed-form expression, except for some special cases, such as Gaussian or exponential pdfs. Alternatively, particle filtering can be applied to approximate $p(x_k|\mathbf{z}_{0:k})$ by stochastic samples generated using a sequential importance sampling strategy.

Since particle filtering with the prior importance function employs no information from observations in proposing new samples, its use is often ineffective and leads to poor filtering performance. Here, we implement a slightly different version of the Gaussian mixture sigma point particle filter (GMSPPF) proposed in [96]; this will be referred to as a composite approach. This composite approach comes out from the utilization of another filtering technique producing a filtering probability density function used as the importance function (IF) for the particle filtering.

The GMSPPF is a family of methodologies that use hybrid sequential Monte-Carlo simulation and a Gaussian sum filter to efficiently estimate posterior distributions of unknown states in a non-linear dynamic system. However, in our state-space modeling, because of the linear model, we modify this method further. Following [96], we will next describe briefly the general framework assumed by the GMKPF method, obtained by replacing the sigma point Kalman filter (SPKF) with a KF. We next outline the main features of the proposed approach. First, we remark that any probability density $p(x)$ can be approximated as closely as desired by a Gaussian mixture model (GMM) of the following form [5]:

$$p(x) \approx p_g(x) = \sum_{g=1}^{G} \alpha^{(g)} N(x; \mu^{(g)}, P^{(g)}) \,, \tag{14.4}$$

where G stands for the number of mixing components, $\alpha^{(g)}$ denotes the mixing weights, and $N(x;\mu,P)$ is a normal distribution with mean μ and covariance P. Thus, the predicted and updated Gaussian components, i.e., the means and covariances of the involved probability densities (posterior, importance, and so on) are calculated using the KF instead of the SPKF [43], [96]. Since the state and observation equations are linear, the KF is employed instead of the SPKF, therefore, the resulting approach is called the Gaussian mixture Kalman particle filter (GMKPF). In order to avoid the particle depletion problem in cases where the observation (measurement) likelihood is very peaked, the GMKPF represents the posterior density by a GMM which is recovered from the resampled equally weighted particle set using the EM algorithm.

In general, for the particle filtering approach, the posterior density $p(x_{0:k}|\mathbf{z}_{1:k})$, where $x_{0:k} = \{x_0, \dots, x_k\}$ and $\mathbf{z}_{1:k} = \{\mathbf{z}_1, \dots, \mathbf{z}_k\}$, constitutes the complete solution to the sequential estimation problem. Our objective is to generate samples from the distribution $p(x_{0:k}|\mathbf{z}_{1:k})$. For this purpose, we collect N sets of samples $x_{0:k}^{(i)} = \{x_0^{(i)}, \dots, x_k^{(i)}\}$ with weights $w_k^{(i)}, i = 1, \dots, N$, and the particles $\{x_k^{(i)}, w_k^{(i)}\}_{i=1}^N$ approximate $p(x_{0:k}|\mathbf{z}_{1:k})$. Finally, the conditional mean state and the corresponding error covariance are calculated as follows:

$$\bar{x}_k = \sum_{i=1}^{N} w_k^{(i)} x_k^{(i)}, \quad \Phi = \sum_{i=1}^{N} w_k^{(i)} [\bar{x}_k - x_k^{(i)}][\bar{x}_k - x_k^{(i)}]^T. \tag{14.5}$$

At the end of each recursion, the particles are resampled to ensure they occur with the same probability as the weights.

The GMKPF combines the IS-based measurement update step with a KF-based Gaussian sum filter for the time update and proposal density generation. In the time update stage, GMKPF approximates the prior, proposal, and posterior density function as GMMs using banks of parallel KFs. The updated mean and covariance of each mixand follow from the KF updates. In the measurement update stage, the GMKPF uses a finite GMM representation of the posterior filtering density

$$p_g(x_k|z_k) = \sum_{g=1}^{G} \alpha_k^{(g)} N(x_k; \mu_k^{(g)}, P_k^{(g)}), \tag{14.6}$$

where G stands for the number of mixing components, $\alpha_l^{(g)}$ are the mixing weights and $N(x_k; \mu_k^{(g)}, P_k^{(g)})$ denotes a normal distribution determined from the gth KF with predicted mean $\mu_k^{(g)} = \bar{x}_k$ and positive definite covariance $P_k^{(g)}$. This is recovered from the weighted posterior particle set of the IS-based measurement update stage, by means of an EM [73] step. The EM algorithm can be used to obtain Gaussian mixture approximations from these particles and weights. Through this mechanism, the EM-based posterior GMM further mitigates the "sample depletion" problem through its inherent "kernel smoothing" nature. The EM algorithm provides an iterative method of estimating $\bar{\theta}$ via

$$\bar{\theta} = \arg\max_{\theta} \; p(z|\theta) \,, \tag{14.7}$$

with the Gaussian mixture specified by the parameter set $\theta = \{\alpha_l^{(1)}, \ldots, \alpha_l^{(G)}, \mu_l^{(1)}, \ldots, \mu_l^{(G)}, P_l^{(1)}, \ldots, P_l^{(G)}\}$. Specifically, the EM algorithm is a two-step iterative algorithm which works as follows: given a $\theta^{(j)}$, it finds the next value $\theta^{(j+1)}$ via

- E step : $Q(\theta|\theta^{(j)}) = E[log\; p(z|\theta)|\theta^{(j)}]$,
- M step : $\theta^{(j+1)} = arg\; \underset{\theta}{max}\; Q(\theta|\theta^{(j)})$.

The reader is referred to [73] for a more detailed explanation of the EM algorithm for GMM. Finally, the conditional mean state estimate and the corresponding error covariance can be calculated as follows:

$$\bar{x}_k = \sum_{g=1}^{G} \alpha_k^{(g)} \mu_k^{(g)}, \quad \bar{P}_k = \sum_{g=1}^{G} \alpha_k^{(g)} [P_k^{(g)} + (\mu_k^{(g)} - \bar{x}_k)(\mu_k^{(g)} - \bar{x}_k)^T]. \tag{14.8}$$

Below we provide a pseudo-code representation for the GMKPF algorithm that, as the next section will illustrate, is very good for estimating clock offsets in non-Gaussian delay models.

GMKPF Algorithm

1. At time k − 1, initialize the densities:
   - the posterior density is approximated by $p_g(x_{k-1}|\mathbf{z_{k-1}}) = \sum_{g=1}^{G} \alpha_{k-1}^{(g)} N(x_{k-1}; \mu_{k-1}^{(g)}, P_{k-1}^{(g)})$;
   - the process noise density is approximated by $p_g(v_{k-1}) = \sum_{i=1}^{I} \beta_{k-1}^{(i)} N(v_{k-1}; \mu_{v_{k-1}}^{(i)}, Q_{k-1}^{(i)})$;
   - the observation noise density is approximated by $p_g(\mathbf{n}_k) = \sum_{j=1}^{J} \gamma_k^{(J)} N(\mathbf{n}_k; \mu_{\mathbf{n}_k}^{(j)}, \mathbf{R}_k^{(j)})$.
2. Preprediction step:
   - calculate the prepredictive state density using KF, $\tilde{p}_g(x_k|\mathbf{z}_{k-1})$;
   - calculate the preposterior state density using KF, $\tilde{p}_g(x_k|\mathbf{z}_k)$.
3. Prediction step:
   - calculate the predictive state density using GMM, $\hat{p}_g(x_k|\mathbf{z}_{k-1})$;
   - calculate the posterior state density using GMM, $\hat{p}_g(x_k|\mathbf{z}_k)$.

4. Observation update step:

   - draw N samples $\{\chi_k^{(l)}; l = 1, \ldots, N\}$ from the importance density function, $q(x_k|\mathbf{z}_k) = \hat{p}_g(x_k|\mathbf{z}_k)$;
   - calculate their corresponding importance weights $\tilde{w}_k^{(l)} = \frac{p(\mathbf{z}_k|\chi_k^{(l)})\hat{p}_g(\chi_k^{(l)}|\mathbf{z}_{k-1})}{\hat{p}_g(\chi_k^{(l)}|\mathbf{z}_k)}$;
   - normalize the weights $w_k^{(l)} = \tilde{w}_k^{(l)} / \sum_{l=1}^{N} \tilde{w}_k^{(l)}$;
   - approximate the state posterior distribution using the EM algorithm, $p_g(x_k|\mathbf{z}_k)$.

5. Infer the conditional mean and covariance:

   - $\bar{x}_k = \sum_{l=1}^{N} w_k^{(l)} \chi_k^{(l)}$ and $\bar{P}_k = \sum_{l=1}^{N} w_k^{(l)} (\chi_k^{(l)} - \bar{x}_k)(\chi_k^{(l)} - \bar{x}_k)^T$
   - or equivalently, upon fitting the posterior GMM, calculate the variables in (14.8).

14.3 Testing the Performance of GMKPF

In this section, computer simulation results will be presented to assess the performance of GMKPF, GML (6.2), and EML (6.18) approaches for estimating the clock offset in WSNs. We consider a total of ten delay models: asymmetric Gaussian, exponential, Gamma, Weibull, and mixtures of Gaussian and exponential, Gaussian and Gamma, Gaussian and Weibull, exponential and Gamma, exponential and Weibull, and Gamma and Weibull distributions. The reason for this comprehensive study is to illustrate that the proposed method is robust, exhibits superior performance, and can be applied to deal with arbitrary network delay distributions. The stationary process v_k is assumed to achieve a given constant variance $Q = 1e - 4$. The numbers of particles and GMM components are 100 and 3, respectively.

Figures 14.1–14.4 show the MSE of the estimators assuming that the random delay models are asymmetric Gaussian, exponential, Gamma, and Weibull pdfs, respectively. The subscripts 1 and 2 are used to differentiate between the parameters of delay distributions corresponding to the uplink and the downlink, respectively. As an example, the parameters σ_1 and σ_2 in Figure 14.1 denote the standard deviations of uplink and downlink asymmetric Gaussian network delay densities, respectively. The MSEs are plotted against the number of observations, which ranges from 5 to 25. Note that the GMKPF performs much better (a reduction in MSE of over 100%) when compared to GML or EML. It is interesting to note that the MSE

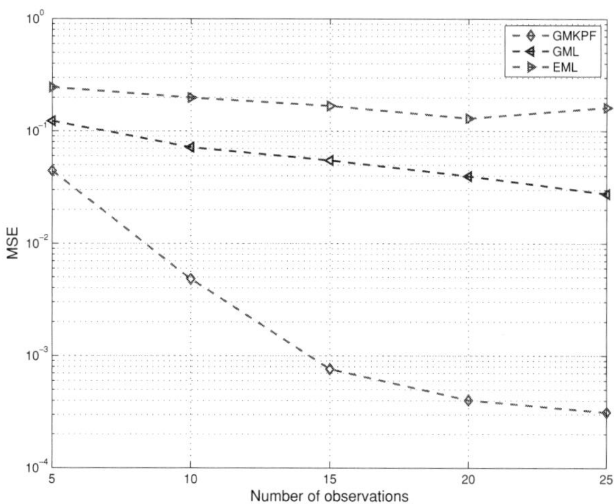

Figure 14.1: MSEs of clock offset estimators for asymmetric Gaussian random delays ($\sigma_1 = 1, \sigma_2 = 4$).

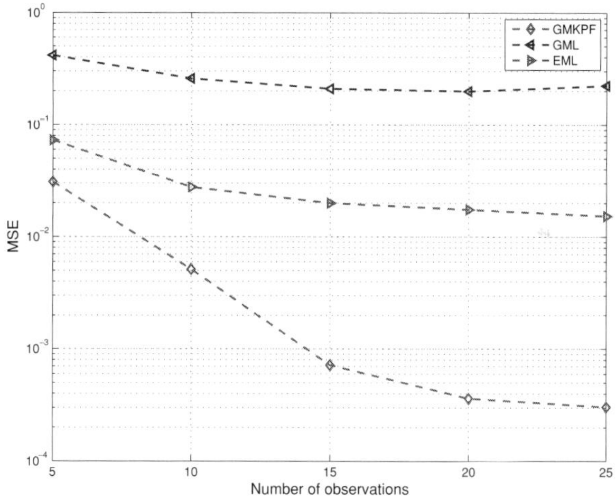

Figure 14.2: MSEs of clock offset estimators for asymmetric exponential random delays ($\lambda_1 = 1, \lambda_2 = 5$).

of GML exhibits better performance than that of EML in the asymmetric Gaussian delay model case and poorer performance in the presence of asymmetric exponential, Gamma, and Weibull delay models. The reason for this is that Gamma and Weibull delay models are closer to the exponential distribution than the Gaussian distribution.

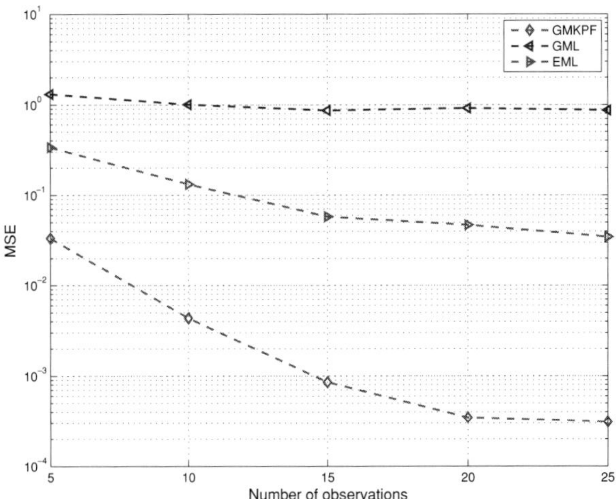

Figure 14.3: MSEs of clock offset estimators for Gamma random delays ($\alpha_1 = 2$, $\beta_1 = 1$).

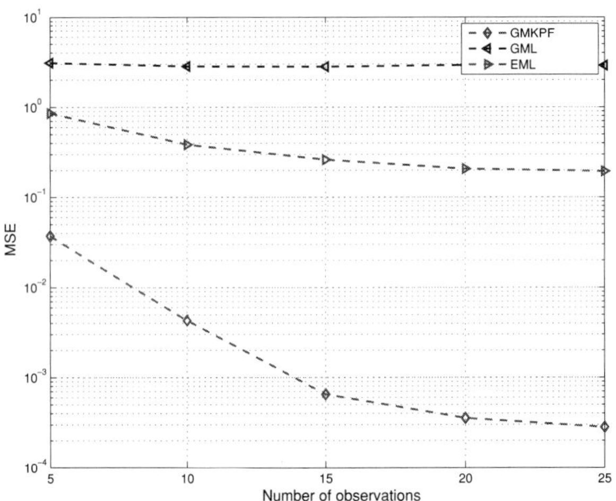

Figure 14.4: MSEs of clock offset estimators for Weibull random delays ($\alpha_1 = 2$, $\beta_1 = 2$) and ($\alpha_2 = 6, \beta_2 = 2$).

To quantify the robustness of the estimators further, we studied the performance of the GMKPF, GML, and EML under various network delay conditions, where the random delay models are mixtures of two distributions. For example, in Figure 14.5, we mix equally a Gaussian mode with an exponential delay model, each having a weight of 50%. This means that if ten observations are received, five observations are

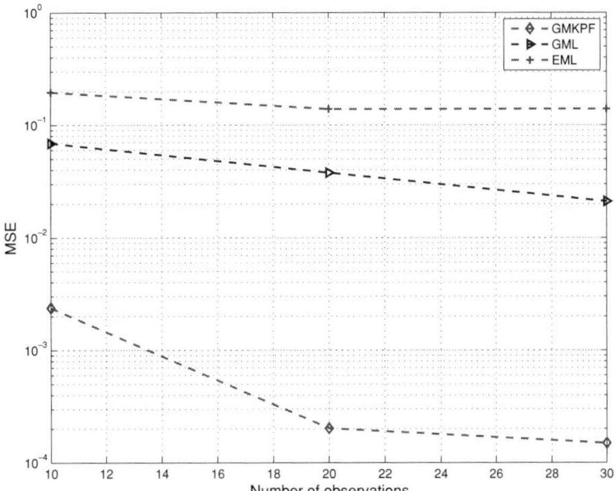

Figure 14.5: MSEs of clock offset estimators for mixing of Gaussian $[\sigma_1 = 1, \sigma_2 = 1]$ and exponential $[\lambda_1 = 1, \lambda_2 = 5]$ random delays.

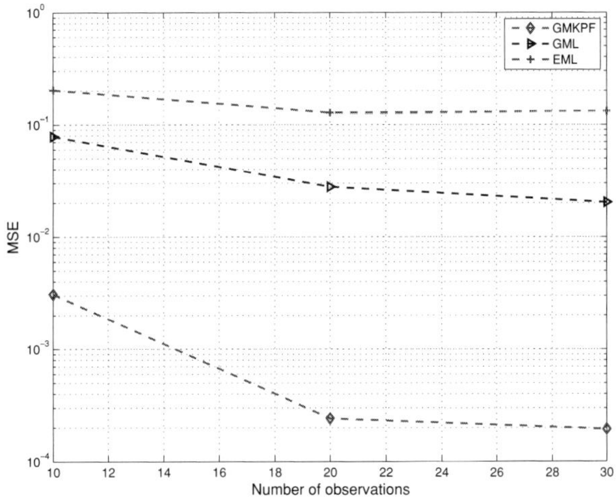

Figure 14.6: MSEs of clock offset estimators for mixing Gaussian $[\sigma_1 = 1, \sigma_2 = 1]$ and Gamma $[\alpha_1 = 2, \beta_1 = 2]$ random delays.

Gaussian and the remaining five samples assume an exponential distribution. From Figures 14.5–14.10, we observe that GMKPF clearly outperforms the GML and EML. In these cases, the GML gives better performance than EML if the network delay process is closer to a Gaussian. Otherwise, the EML exhibits better performance than the GML, while GMKPF outperforms both the GML and EML.

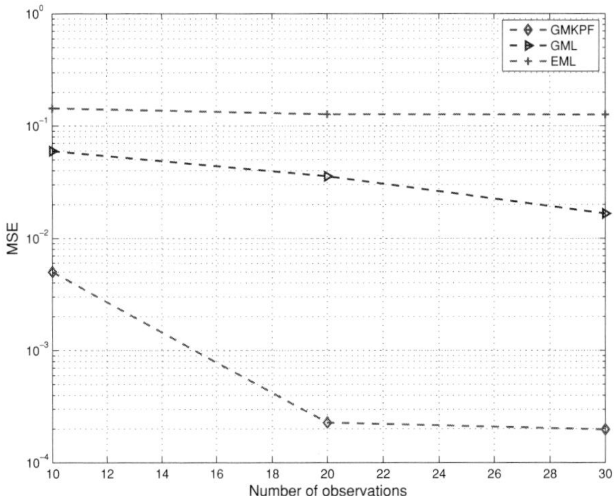

Figure 14.7: MSEs of clock offset estimators for mixing Gaussian $[\sigma_1 = 1, \sigma_2 = 1]$ and Weibull $[\alpha_1 = 2, \beta_1 = 2$ and $\alpha_2 = 6, \beta_2 = 2]$ random delays.

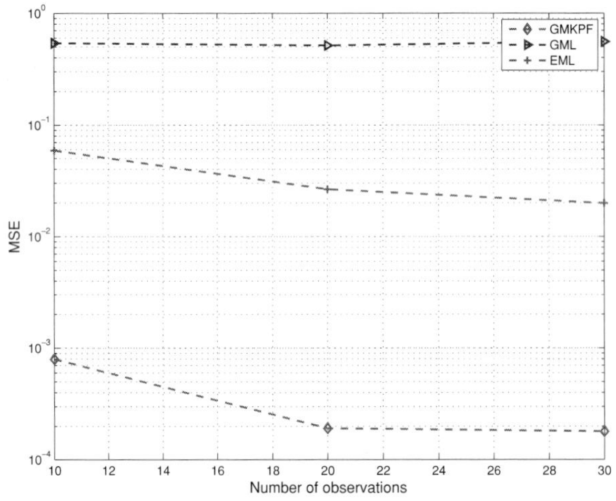

Figure 14.8: MSEs of clock offset estimators for mixing exponential $[\lambda_1 = 1, \lambda_2 = 5]$ and Gamma $[\alpha_1 = 2, \beta_1 = 5]$, $[\alpha_2 = 2, \beta_2 = 2]$ random delays.

14.4 COMPOSITE PARTICLE FILTERING (CPF) WITH BOOTSTRAP SAMPLING (BS)

Under the Bayesian framework, an emerging powerful technique for obtaining the posterior, predictive, and filtering pdf is referred to as particle filtering

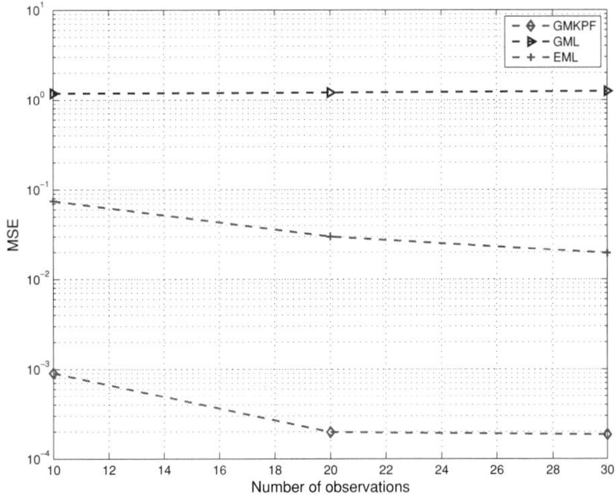

Figure 14.9: MSEs of clock offset estimators for mixing exponential $[\lambda_1 = 1, \lambda_2 = 5]$ and Weibull $[\alpha_1 = 2, \beta_1 = 2$ and $\alpha_2 = 6, \beta_2 = 2]$ random delays.

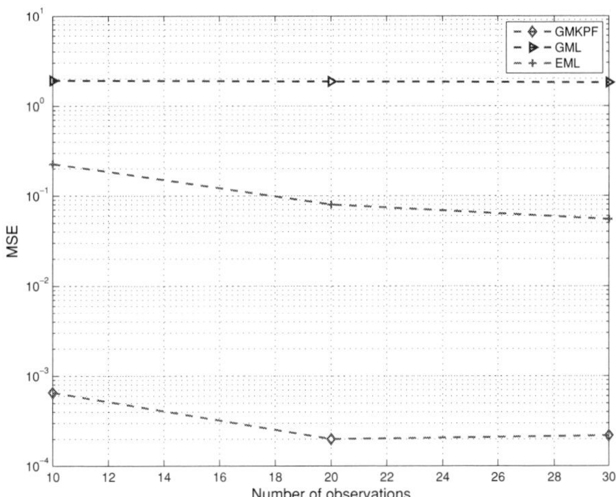

Figure 14.10: MSEs of clock offset estimators for mixing Gamma with parameters $[\alpha_1 = 2, \beta_1 = 5]$ and $[\alpha_2 = 2, \beta_2 = 2]$ and Weibull $[\alpha_1 = 2, \beta_1 = 2$ and $\alpha_2 = 6, \beta_2 = 2]$ random delays.

(see e.g., [6], [18]). The particle filtering technique allows a complete representation of the state posterior distribution, which approximates $p(x_k|\mathbf{z}_{0:k})$, by stochastic samples generated using a sequential IS strategy. The most common particle filtering strategy is to sample from the transition prior distribution due to its simplicity. Since the prior IS distribution employs no information from observations in proposing new samples, its use is often ineffective and leads to poor filtering performance. To overcome these challenges, we will next derive an extension of GSPF [48] applicable for linear non-Gaussian models.

In GSPF, we would like to obtain the filtering and predictive distributions recursively represented as finite Gaussian mixtures using Gaussian particle filtering (GPF) [47]. One set of methods approximates the mixture components of the predictive and filtering distributions as Gaussian. The approximation can be implemented by the GSPF, resulting in a parallel bank of GPFs. Since Gaussian mixture models are increasingly used for modeling non-Gaussian densities [4], [77], here we extend the use of the GSPF to linear non-Gaussian models. The resulting new approach will be referred to as CPF. Notice that in the measurement and time-update equations of CPF, the updated mean and covariance of each mixand follow from the KF. The CPF is implemented by means of G parallel KFs, and the weights are adjusted according to the given update equations. Notice also that the CPF approach comes from the utilization of another filtering technique (KF) producing a filtering pdf used as the IF for the particle filtering.

We next describe the general framework of CPF methods. Assume at time $k-1$, the posterior distribution $p(x_{k-1}|\mathbf{z}_{0:k-1})$ is approximated as closely as desired by a GMM of the following form [5]:

$$p(x_{k-1}|\mathbf{z}_{0:k-1}) \approx \sum_{g=1}^{G} w_{(k-1)g} N(x_{k-1}; \mu_{(k-1)g}, P_{(k-1)g}), \tag{14.9}$$

where G stands for the number of mixing components, $w_{(k-1)g}$ denotes the mixing weights, and $N(x;\mu,P)$ represents the normal distribution of RV x with mean μ and covariance P. The transition prior is given by

$$p(x_k|x_{k-1}) = \sum_{l=1}^{L} \alpha_{kl} N(x_k; Fx_{k-1} + \bar{\mu}_{(k-1)l}, \bar{P}_{(k-1)l}). \tag{14.10}$$

The time-update stage in which the previous observations and state are used to predict the current state is encompassed by the predictive distribution which can be approximated as follows:

$$\begin{aligned} p(x_k|\mathbf{z}_{1:k-1}) &= \int p(x_k|x_{k-1}) p(x_{k-1}|\mathbf{z}_{1:k-1}) dx_{k-1} \\ &= \int p(x_k|x_{k-1}) \sum_{g=1}^{G} w_{(k-1)g} N(x_{k-1}; \mu_{(k-1)g}, P_{(k-1)g}) dx_{k-1} \end{aligned}$$

$$= \int \sum_{l=1}^{L} \alpha_{kl} N(x_k; Fx_{k-1} + \bar{\mu}_{(k-1)l}, \bar{P}_{(k-1)l}) \times \sum_{g=1}^{G} w_{(k-1)g} N(x_{k-1}; \mu_{(k-1)g}, P_{(k-1)g}) dx_{k-1}. \quad (14.11)$$

The predicted and updated Gaussian component means and covariances are calculated using the KF. As in the KF, the integral in (14.11) is approximated by a Gaussian. Then, the predictive distribution can be approximated as

$$p(x_k|\mathbf{z}_{1:k-1}) \approx \sum_{j=1}^{GL} \tilde{w}_{kg} N(x_k; \tilde{\mu}_{kj}, \tilde{P}_{kj}), \quad (14.12)$$

where the parameters of the mixture are obtained according to KF:

$$\begin{aligned} \tilde{\mu}_{kj} &= \bar{\mu}_{(k-1)l} + F\mu_{(k-1)g}, & (14.13)\\ \tilde{P}_{kj} &= FP_{(k-1)g}F^T + \bar{P}_{(k-1)l}, & (14.14)\\ \tilde{w}_{kj} &= \alpha_{kl} w_{(k-1)g}, & (14.15) \end{aligned}$$

for appropriate $j = 1, \ldots, GL$, $g = 1, \ldots, G$ and $l = 1, \ldots, L$ and $j = g + (l-1)G$.

In the measurement update stage, the operation of acting on new observations to improve on previously "predictive" states $p(x_k|\mathbf{z}_{1:k-1})$ can be approximated as

$$\begin{aligned} p(x_k|\mathbf{z}_{1:k}) &= \frac{p(\mathbf{z}_k|x_k)p(x_k|\mathbf{z}_{1:k-1})}{p(\mathbf{z}_k|\mathbf{z}_{k-1})} \\ &\approx C_k \sum_{j=1}^{GL} \tilde{w}_{kj} p(\mathbf{z}_k|x_k) N(x_k; \tilde{\mu}_{kj}, \tilde{P}_{kj}), \quad (14.16) \end{aligned}$$

where C_k is the normalizing constant. Therefore, the updated filtering distribution is approximated as

$$p(x_k|\mathbf{z}_{1:k}) \approx \sum_{j=1}^{GL} w_{kj} N(x_k; \mu_{kj}, P_{kj}). \quad (14.17)$$

The reader is referred to [48] for more detailed explanations of the GSPF algorithm. Finally, notice that the conditional mean state estimate and the corresponding error covariance can be calculated respectively as:

$$\bar{x}_k = \sum_{j=1}^{GL} w_{kj}\, \mu_{kj}, \quad \bar{P}_k = \sum_{j=1}^{GL} w_{kj} \left[P_{kj} + (\mu_{kj} - \bar{x}_k)(\mu_{kj} - \bar{x}_k)^T\right]. \quad (14.18)$$

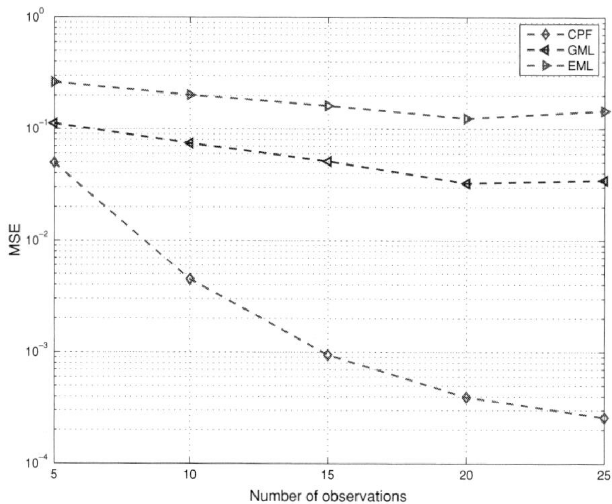

Figure 14.11: MSEs of clock offset estimators for asymmetric Gaussian random delays $[\sigma_1 = 1, \sigma_2 = 4]$.

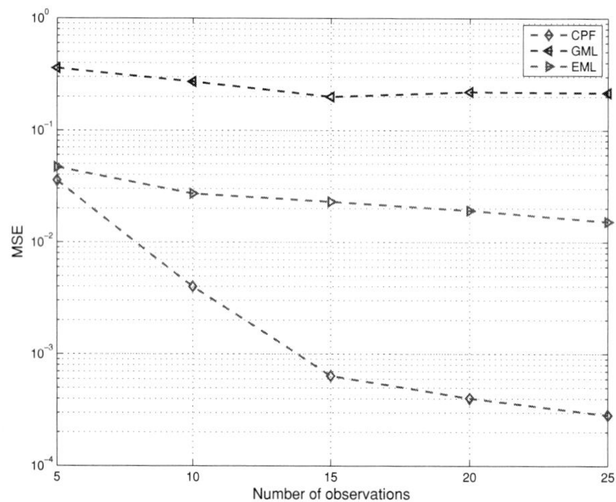

Figure 14.12: MSEs of clock offset estimators for asymmetric exponential random delays $[\lambda_1 = 1, \lambda_2 = 5]$.

Next, we introduce another clock estimation scheme obtained through the integration of the CPF technique with the BS approach. The reader is referred to [21], [101] for more detailed explanations of BS. In order to provide a consistent number of observation data in the presence of errors during timing message transmissions, new sampled observation data are generated from the original observation data via

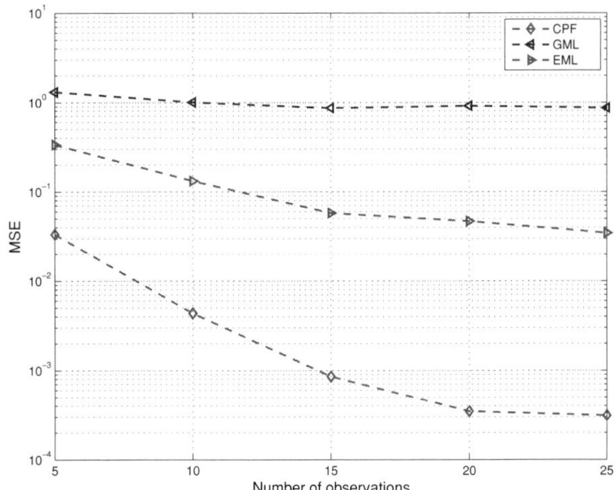

Figure 14.13: MSEs of clock offset estimators for Gamma random delays [$\alpha_1 = 2$, $\beta_1 = 1$].

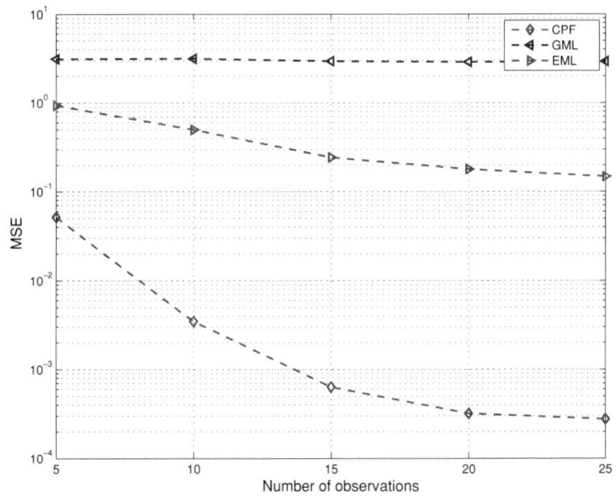

Figure 14.14: MSEs of clock offset estimators for Weibull random delays [$\alpha_1 = 2$, $\beta_1 = 2$ and $\alpha_2 = 6, \beta_2 = 2$].

the BS. Then, the clock offset is estimated based on the CPF. Notice that even in the presence of corrupted or lost data packets, BS can create additional samples from the original sample set, by drawing replacements randomly from **Z**, without the necessity of additional retransmissions. Each of the bootstrap samples is considered as a new datum. Based on the additional sampled observation data, we can then approximate the clock offset x by using the CPF. The major steps

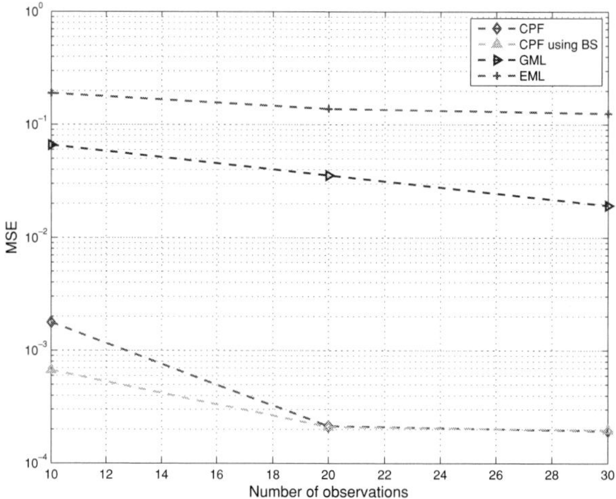

Figure 14.15: MSEs of clock offset estimators for a mixture of Gaussian [$\sigma_1 = 1$, $\sigma_2 = 1$] and exponential [$\lambda_1 = 1, \lambda_2 = 5$] random delays.

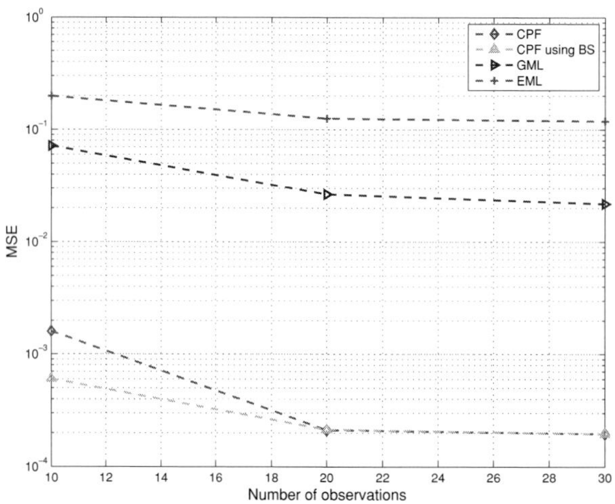

Figure 14.16: MSEs of clock offset estimators for a mixture of Gaussian [$\sigma_1 = 1$, $\sigma_2 = 1$] and Gamma [$\alpha_1 = 2, \beta_1 = 2$] random delays.

of the CPF approach with bootstrap sampling are summarized by the following pseudo-code:

Algorithm: CPF with BS

1. Conduct the experiment to obtain the random sample $\mathbf{Z} = \{\mathbf{Z}_1, \ldots, \mathbf{Z}_n\}$ and calculate the estimate $\hat{\theta}$ from the sample $\mathbf{Z}$.

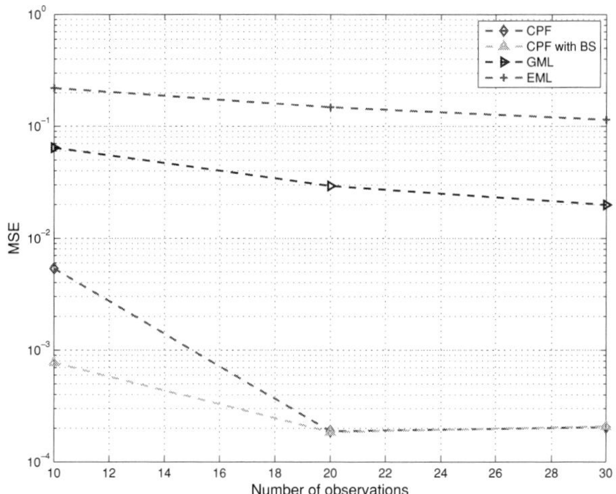

Figure 14.17: MSEs of clock offset estimators for a mixture of Gaussian [$\sigma_1 = 1$, $\sigma_2 = 1$] and Weibull [$\alpha_1 = 2, \beta_1 = 2$ and $\alpha_2 = 6, \beta_2 = 2$] random delays.

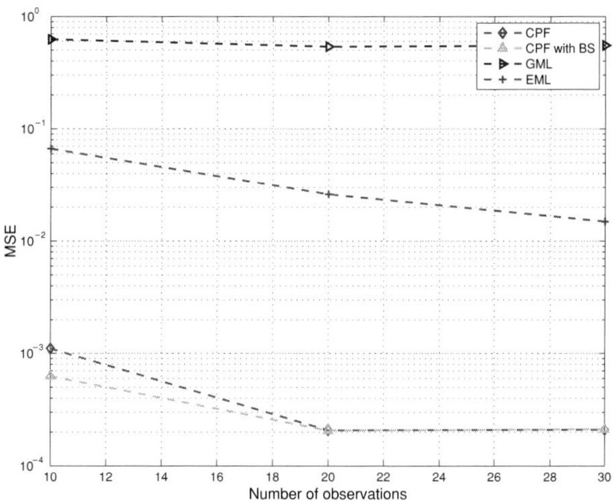

Figure 14.18: MSEs of clock offset estimators for a mixture of exponential [$\lambda_1 = 1$, $\lambda_2 = 5$] and Gamma [$\alpha_1 = 2, \beta_1 = 5$ and $\alpha_2 = 2, \beta_2 = 2$] random delays.

2. Construct the empirical distribution $\hat{H}$, which gives equal weight, $1/n$, to each observation $\mathbf{Z}_1 = \mathbf{z}_1, \ldots, \mathbf{Z}_n = \mathbf{z}_n$.

3. From $\hat{H}$, draw a sample $\mathbf{Z}^* = \{\mathbf{Z}_1^*, \ldots, \mathbf{Z}_n^*\}$, an operation called bootstrap resampling.

4. From the bootstrap resample $\mathbf{Z}^*$, estimate the clock offset $\hat{x}$ by CPF.

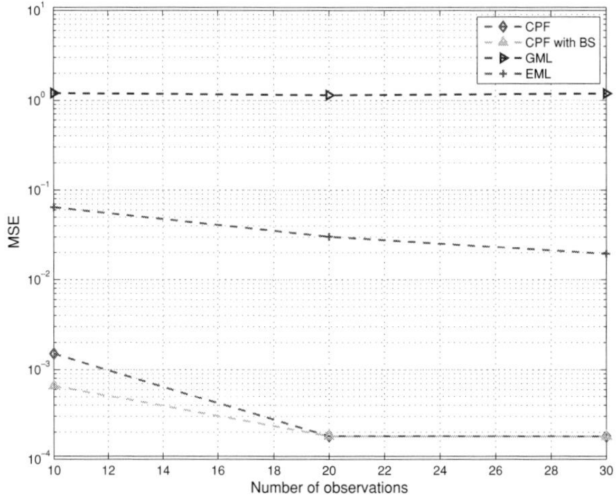

Figure 14.19: MSEs of clock offset estimators for a mixture of exponential [$\lambda_1 = 1$, $\lambda_2 = 5$] and Weibull [$\alpha_1 = 2, \beta_1 = 2$ and $\alpha_2 = 6, \beta_2 = 2$] random delays.

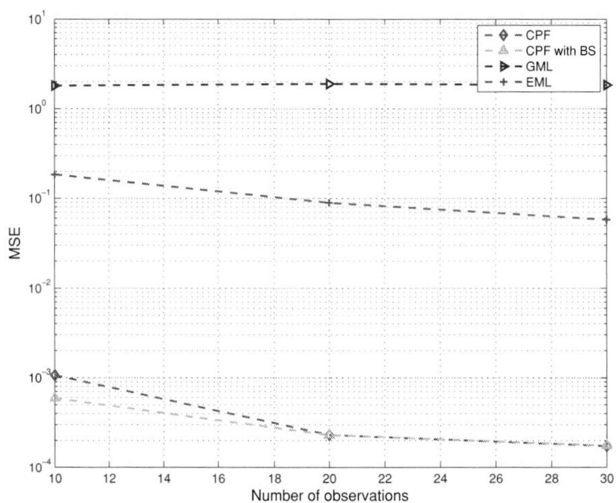

Figure 14.20: MSEs of clock offset estimators for a mixture of Gamma [$\alpha_1 = 2, \beta_1 = 5$ and $\alpha_2 = 2, \beta_2 = 2$] and Weibull [$\alpha_1 = 2, \beta_1 = 2$ and $\alpha_2 = 6, \beta_2 = 2$] random delays.

14.5 TESTING THE PERFORMANCE OF CPF AND CPF WITH BS

In this section, extensive computer simulation results are presented to illustrate the performance of the CPF, CPF with BS, GML (6.2), and EML (6.18) approaches for estimating the clock offset in WSNs, assuming a variety of random network delay models, such as asymmetric Gaussian, exponential, Gamma, and Weibull as

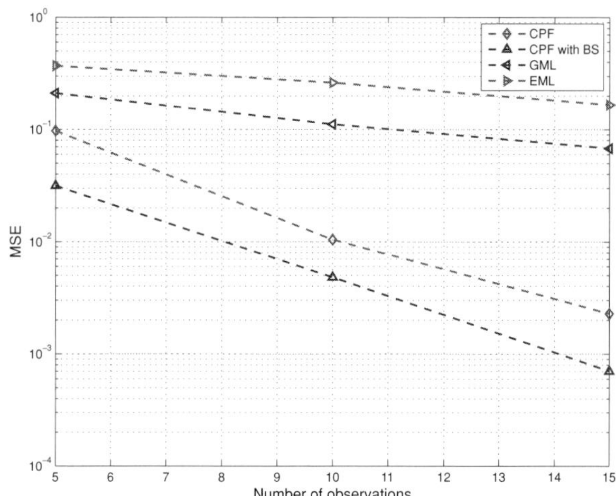

Figure 14.21: MSEs of clock offset estimators for asymmetric Gaussian random delay $[\sigma_1 = 1, \sigma_2 = 4]$ and two message exchange errors.

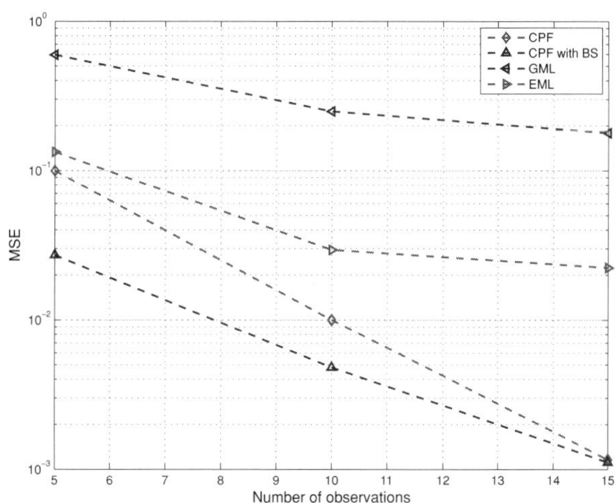

Figure 14.22: MSEs of clock offset estimators for asymmetric exponential random delay $[\lambda_1 = 1, \lambda_2 = 5]$ and two exchange message errors occur.

well as mixtures of Gaussian and exponential, Gaussian and Gamma, Gaussian and Weibull, exponential and Gamma, exponential and Weibull, and Gamma and Weibull. This rather extensive set of simulations is conducted to illustrate the fact that the proposed method can be widely and flexibly applied for general arbitrary distributions. The stationary process v_k assumes the constant variance $Q = 1e - 4$, while the numbers of particles and GMM components are set to 100 and 3, respectively.

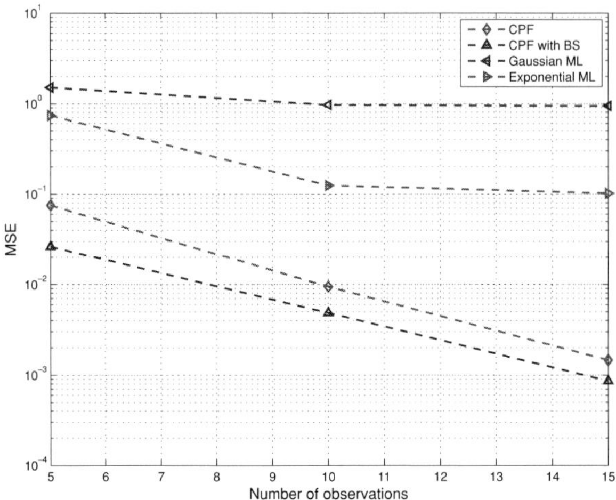

Figure 14.23: MSEs of clock offset estimators for Gamma random delay [$\alpha_1 = 2$, $\beta_1 = 1$] and two exchange message errors occur.

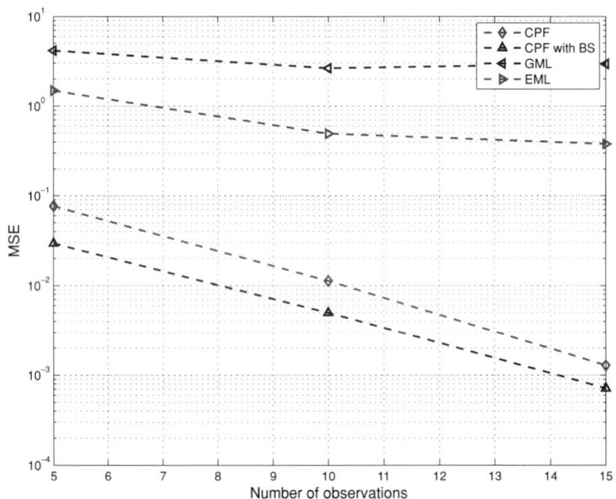

Figure 14.24: MSEs of clock offset estimators for Weibull random delay [$\alpha_1 = 2$, $\beta_1 = 2$ and $\alpha_2 = 6, \beta_2 = 2$] and two exchange message errors occur.

Figures 14.11–14.14 show the MSE of the estimators in conditions where the network delay distributions are asymmetric Gaussian, exponential, Gamma, and Weibull pdfs, respectively. The subscripts attached to the distribution parameters are used to differentiate between the parameters of the uplink distribution and those of the downlink distribution. For example, for an asymmetric Gaussian delay model, (σ_1^2) and (σ_2^2) denote the uplink and downlink variances of network delays, respectively. The MSE curves are plotted against the number of observations

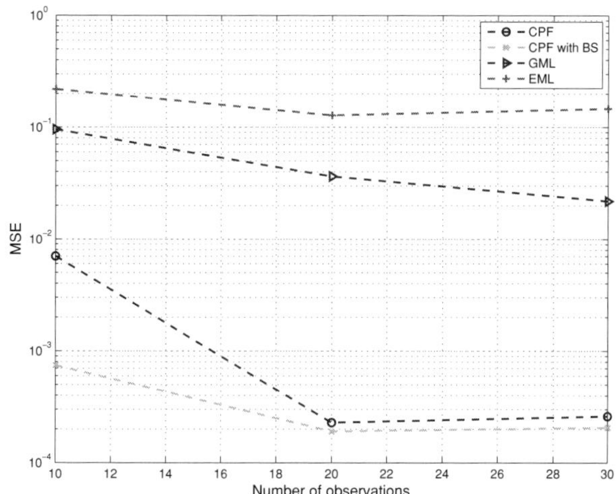

Figure 14.25: MSEs of clock offset estimators for a mixture of Gaussian $[\sigma_1 = 1, \sigma_2 = 1]$ and exponential $[\lambda_1 = 1, \lambda_2 = 5]$ and two exchange message errors occur.

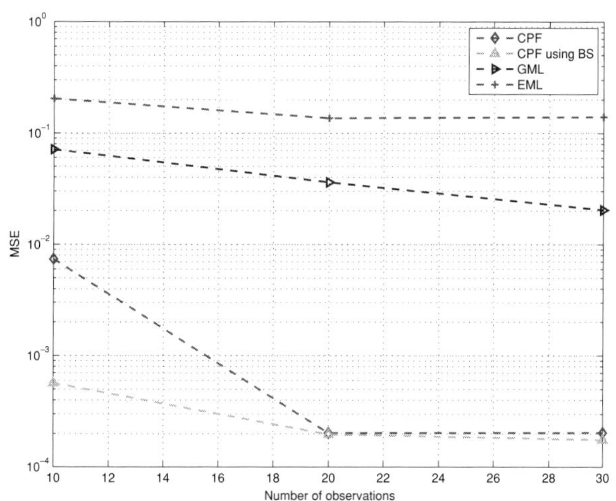

Figure 14.26: MSEs of clock offset estimators for a mixture of Gaussian $[\sigma_1 = 1, \sigma_2 = 1]$ and Gamma $[\alpha_1 = 2, \beta_1 = 2]$ random delays and two exchange message errors occur.

which range from 5 to 25. Note that the CPF performs much better with over 100% MSE reduction when compared to the GML or EML. Notice also that the CPF with BS exhibits the best performance when there is a reduced number of observation data. Notice also the MSE of GML achieves better performance than EML in asymmetric Gaussian delay models, while EML has a superior performance relative to

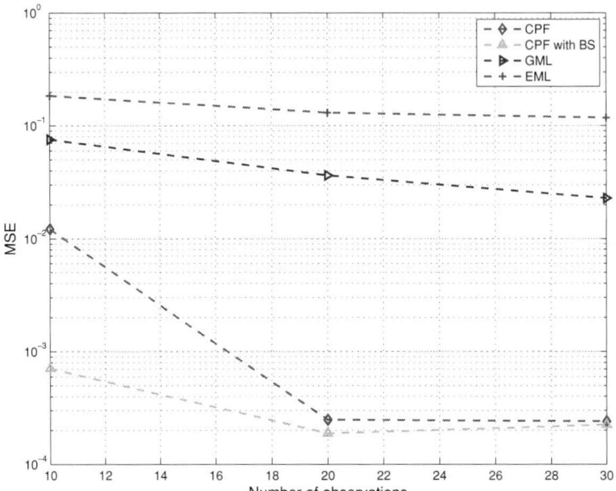

Figure 14.27: MSEs of clock offset estimators for mixing of Gaussian $[\sigma_1 = 1, \sigma_2 = 1]$ and Weibull $[\alpha_1 = 2, \beta_1 = 2$ and $\alpha_2 = 6, \beta_2 = 2]$ random delays and two exchange message errors occur.

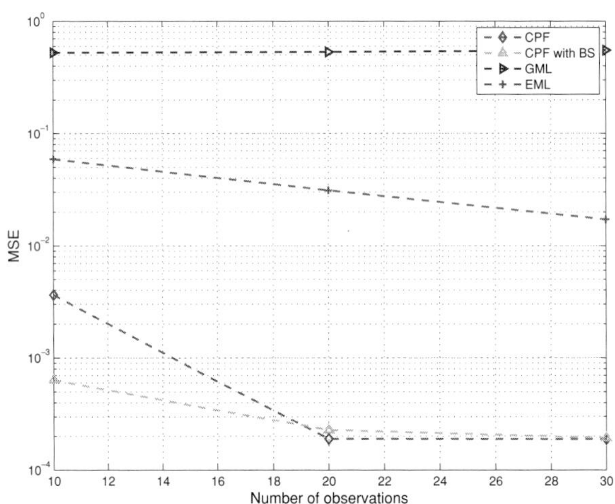

Figure 14.28: MSEs of clock offset estimators for a mixture of exponential $[\lambda_1 = 1, \lambda_2 = 5]$ and Gamma $[\alpha_1 = 2, \beta_1 = 5$ and $\alpha_2 = 2, \beta_2 = 2]$ random delays and two exchange message errors.

GML in asymmetric exponential, Gamma, and Weibull delay models. The reason for this is that Gamma and Weibull delay models are closer to the exponential distribution than Gaussian.

To quantify the robustness of the estimators further, we studied the performance of the CPF with BS, CPF, GML, and EML under various conditions, where the

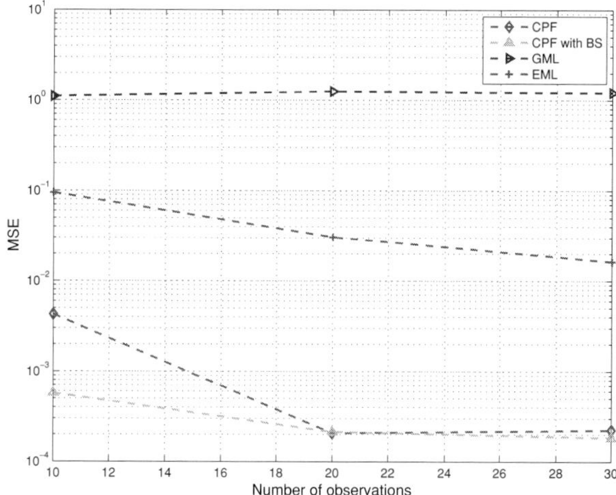

Figure 14.29: MSEs of clock offset estimators for a mixture of exponential $[\lambda_1 = 1, \lambda_2 = 5]$ and Weibull $[\alpha_1 = 2, \beta_1 = 2$ and $\alpha_2 = 6, \beta_2 = 2]$ random delays and two exchange message errors.

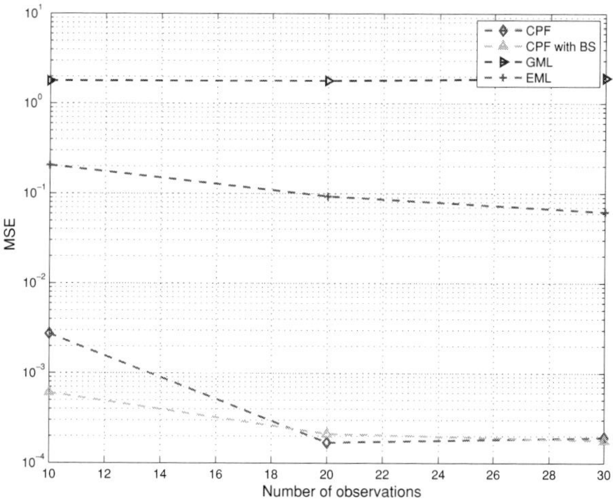

Figure 14.30: MSEs of clock offset estimators for a mixture of Gamma $[\alpha_1 = 2, \beta_1 = 5$ and $\alpha_2 = 2, \beta_2 = 2]$ and Weibull $[\alpha_1 = 2, \beta_1 = 2$ and $\alpha_2 = 6, \beta_2 = 2]$ random delays and two exchange message errors.

random delay models are mixed. For example, in Figure 14.15, we mixed the Gaussian and the exponential delay model uniformly, each distribution accounting for 50% of samples. This means that if ten observations are recorded, five are Gaussian and five are exponentially distributed. From Figures 14.15–14.20, we observe that

CPF clearly outperforms the GML and EML. In these cases, the GML also gives better performance than EML when the network random delay distribution is closer to a Gaussian. Notice that CPF outperforms both the GML and EML no matter what distribution model is assumed for the network delays.

Figures 14.21–14.30 depict the MSEs versus the number of observation data when two message exchange errors occur with uniform distribution for the scenarios assumed by Figures 14.11–14.20. From the Figures 14.21–14.24, we observe that CPF with BS clearly outperforms CPF, GML, and EML. From Figures 14.26–14.30, we observe that unlike in the Figures 14.21–14.24, CPF with BS exhibits the best performance when there is a reduced number of observations. From these simulation results, it follows that CPF with BS and CPF are reliable methods when there is a reduced number of samples.

Chapter 15

CONCLUSIONS AND FUTURE DIRECTIONS

Much attention has been paid to WSNs due to their capability of serving a variety of purposes. Time synchronization is a significant factor in WSNs, and a number of fundamental operations, such as data fusion, power management, and transmission scheduling, require accurate time synchronization. Since the conventional time synchronization protocol for the Internet cannot be directly applied to WSNs, a number of synchronization protocols have been proposed to meet the unique requirements of sensor network applications.

The importance of time synchronization also comes from the evolution of WSNs which has been driven by technological advances in diverse areas. For instance, unlike the currently deployed WSNs, the next generation of sensor networks may consist of dynamic mobile sensors or a mixture of static and dynamic sensors. In this scenario, far more sophisticated time synchronization protocols that efficiently deal with the mobility of sensors will be required. Indeed, as the sensor network becomes more complicated, the role of time synchronization will become much more important.

In this book, the basic features and theoretical background of the time synchronization problem in WSNs were introduced and then the basic approaches were analyzed and compared to reveal the general ideas and features of time synchronization protocols in WSNs. In addition, a survey of existing time synchronization protocols in the literature was provided including the most recent results.

As a main feature of this book, the problem of time synchronization was studied from a statistical signal processing point of view. This book targeted the clock synchronization problem in a general sender–receiver and receiver–receiver timing packet exchange scenario. The best linear unbiased estimate (using order statistics) of the clock offset between two nodes was derived for both symmetric and asymmetric exponential link delay circumstances. The MVUE was also obtained, and compared with the ML estimator to find the optimal estimator for various regions of asymmetry. Then the MLEs of both the clock offset and skew for any general time synchronization protocol involving a two-way message exchange mechanism were derived assuming both Gaussian and exponential delays, and the algorithms used for finding these MLEs were also detailed. By sacrificing some performance, two reduced complexity schemes were also proposed to estimate the clock parameters in an economical way. The PBS was introduced and shown to require far fewer timing messages (energy consumption) than other well-known schemes without incurring any loss in synchronization accuracy. The ML, best linear unbiased, minimum variance unbiased and MMSE estimators were then derived for the inactive nodes overhearing a timing exchange between a pair of nodes. Based on the SRS approach, the AMTS protocol was introduced to minimize the overall network-wide energy consumption required for global synchronization. Afterwards, a procedure for estimating the clock drift was also presented to provide increased accuracy and a longer synchronization period. Next, the JML estimator was derived in a general receiver–receiver setting and the Gibbs sampler was applied to increase the accuracy.

Chapter 14 proposed several novel methods such as GMKPF, CPF, and CPF with BS, for robust estimation of the clock offset in WSNs in the presence of unknown network delay distributions. The benefits of these methods come in terms of improved performance and applicability to any random delay models, such as asymmetric Gaussian, exponential, Gamma, Weibull, as well as mixtures of these delay models. In addition, the proposed CPF approaches are robust when there are a small number of observations, message exchange errors and unknown network delay distributions. Also, it turns out that CPF with BS achieves excellent performance compared to GML and EML in environments which manifest message exchange errors and time-varying network delay distributions.

The synchronization schemes and theoretical analysis presented in this book are useful for developing (or selecting) more efficient synchronization protocols tailored specifically to the needs of the particular sensor network applications.

There are several future research directions that present both theoretical and practical interest. First, re-examination of the clock offset estimation problems utilizing modern statistical inference techniques such as bootstrap and jackknife is an area worth exploring. The methodology adopted here can be extended to analyze other time synchronization protocols for both single-hop and multi-hop cases. In addition, finding the Cramer–Rao and more general lower bounds (e.g., Chapman–Robinson or Barankin bounds) for the clock offset and skew estimators derived here represent important open research problems. Additionally, exploring the effects of violating the iid assumption for random delays, missing data points due to communication losses, channel fading, and quantization errors are interesting open problems. Moreover, formulating a procedure through which the timing error accumulation over a series of hops encompassing the whole network could be quantified is an idea worth exploring.

In addition, experimental performance evaluation and comparisons with other synchronization protocols are major open research topics. More general random delay models might be needed for some sensor network applications. For instance, the Gamma distributed delay model might be a better choice than the exponential delay model in some cases due to its superior precision, given the degree of freedom offered by an extra free parameter.

The facts that analytical closed-form expressions do not seem to exist for the MSEs of the clock estimators and the computation of lower bound performance bounds appears difficult due to the non-Gaussian nature of involved distributions are open research problems for the CPF-based approaches.

Finally, developing optimal clock offset estimation schemes for global synchronization of large-scale WSNs appears to be a very important and challenging problem. In this regard, investigating factor graph (or message passing) algorithms in combination with distributed and sequential Bayesian, MCMC techniques, and distributed non-linear optimization techniques might hold the answer to developing efficient synchronization protocols for large-scale WSNs.

ACRONYMS

ACK	Acknowledgement Packet
ADP	Asynchronous Diffusion Protocol
AMTS	Adaptive Multi-hop Timing Synchronization
AO	Always On
BLUE	Best Linear Unbiased Estimation
BLUE-OS	Best Linear Unbiased Estimation using Order Statistics
BS	Bootstrap Sampling
CDF	Cumulative Distribution Function
CENS	Center for Embedded Networked Sensing
CLT	Central Limit Theorem
CPF	Composite Particle Filter
CRB	Cramer–Rao Bound
CRLB	Cramer–Rao Lower Bound
DMTS	Delay Measurement Time Synchronization
EM	Expectation–Maximization
EML	Exponential Maximum Likelihood
EMLE	Exponential Maximum Likelihood Estimation
EMLLE	Exponential-Maximum-Likelihood-Like Estimation
FTSP	Flooding Time Synchronization Protocol

GMKPF	Gaussian Mixture Kalman Particle Filter
GML	Gaussian Maximum Likelihood
GMLE	Gaussian Maximum Likelihood Estimation
GMLLE	Gaussian-Maximum-Likelihood-Like Estimation
GMM	Gaussian Mixture Model
GMSPPF	Gaussian Mixture Sigma Point Particle Filter
GPA	Group-wise Pair selection Algorithm
GPF	Gaussian Particle Filter
GPS	Global Positioning System
GSPF	Gaussian Sum Particle Filter
ID	Identifier
IF	Importance Function
iid	independent and identically distributed
IP	Internet Protocol
IS	Importance Sampling
JML	Joint Maximum Likelihood
JMLE	Joint Maximum Likelihood Estimation
KF	Kalman Filter
LDP	Level Discovery Packet
LTS	Lightweight Time Synchronization
M/M/I	Markov single-server queue model
MAC	Medium Access Control
MCMC	Markov Chain Monte-Carlo
MEMS	Micro-Electro-Mechanical System
ML	Maximum Likelihood

MLE	Maximum Likelihood Estimation
MLLE	Maximum-Likelihood-Like Estimation
MMSE	Minimum Mean Square Error
MSE	Mean Square Error
MVU	Minimum Variance Unbiased
MVUE	Minimum Variance Unbiased Estimation
NPA	Network-wide Pair selection Algorithm
NTP	Network Time Protocol
OS	Order Statistics
PBS	Pairwise Broadcast Synchronization
PCO	Pulse Coupled Oscillator
pdf	probability density function
PS	Pairwise Synchronization
RATS	Rate-Adaptive Time Synchronization
RBS	Reference Broadcast Synchronization
RF	Radio Frequency
ROS	Receive-Only Synchronization
RRS	Receiver–Receiver Synchronization
RV	Random Variable
SI	Sensor Initiated
SMA	Sequential Multi-hop synchronization Algorithm
SNR	Signal-to-Noise Ratio
SPKF	Sigma Point Kalman Filter
SRS	Sender–Receiver Synchronization
TDMA	Time Division Multiple Access

TDP Time Diffusion Protocol

TPSN Timing Synch Protocol for Sensor Networks

TSHL Timing Synchronization protocol for High Latency

TTP Time Transmission Protocol

UDP User Datagram Protocol

WSN Wireless Sensor Network

REFERENCES

[1] H. S. ABDEL-GHAFFAR, Analysis of synchronization algorithm with time-out control over networks with exponentially symmetric delays, *IEEE Transactions on Communications*, **50** (2002), 1652–1661.

[2] I. F. AKYILDIZ, W. SU, Y. SANKARASUBRAMANIAM, AND E. CAYIRCI, Wireless sensor networks: a survey, *Computer Networks*, **38** (2002), 393–422.

[3] D. ALLAN, Time and frequency (time-domain) characterization, estimation, and prediction of precision clocks and oscillators, *IEEE Transactions on Ultrasonics, Ferroelectrics, and Frequency Control*, **34** (1987), 647–654.

[4] D. L. ALSPACH AND H. W. SORENSON, Nonlinear bayesian estimation using gaussian sum approximation, *IEEE Transactions on Automatic Contr.*, **AC-17** (1972), 439–448.

[5] B. D. ANDERSON AND J. B. MOORE, *Optimal Filtering*, Prentice-Hall, 1979.

[6] M. SANJEEV ARULAMPALAM, S. MASKEIL, N. GORDON ET. AL., A tutorial on particle filters for on-line non-linear/non-gaussian bayesian tracking, *IEEE Transactions on Signal Processing*, **50** (2002), 174–188.

[7] K. ARVIND, Probabilistic clock synchronization in distributed systems, *IEEE Transactions on Parallel and Distributed Systems*, **5** (1994), 474–487.

[8] B. AWERBUCH AND R. G. GALLAGER, A new distributed algorithm to find breath first search trees, *IEEE Transactions on Information Theory*, **33** (1987), 315–322.

[9] R. R. BAHADUR, Sufficiency and statistical decision functions, *The Annals of Mathematical Statistics*, **25** (1954), 423–462.

[10] J. BESAG, Spatial interaction and the statistical analysis of lattice systems, *Journal of the Royal Statistical Society Series B (Methodological)*, **36** (1974), 192–236.

[11] S. BIAZ AND J. L. WELCH, Closed form bounds for clock synchronization under simple uncertainty assumptions, *Information Processing Letters*, **80** (2001), 151–157.

[12] A. BOUKERCHE AND D. TURGUT, Secure time synchronization protocols for wireless sensor networks, *IEEE Wireless Communications*, **14** (2007), 64–69.

[13] C. BOVY, H. MERTODIMEDJO, G. HOOGHIEMSTRA, H. UIJTERWAAL, AND P. MIEGHEM, Analysis of end-to-end delay measurements in Internet, in the *Passive and Active Measurements Workshop, Fort Collins, CO, Mar 2002*, pp. 26–33. Agilent Technologies, 2002.

[14] N. BULUSU AND S. JHA, *Wireless Sensor Networks: A Systems Perspective*. Artech House, 2005.

[15] CENS, *Habitat sensing group at James Reserve*. `http://www.jamesreserve.edu/`.

[16] F. CRISTIAN, Distributed counting, in *Probabilistic Clock Synchronization*, pp. 146–158. Springer Verlag, 1989.

[17] S. DIXIT AND P. SKELLY, Mpeg-2 over atm or video dialtone networks: issues and strategies, *IEEE Network, Special Issue on Digital Interactive Broadband Video Dial Tone Networks*, **9** (1995), 30–40.

[18] P. M. DJURIC, J. H. KOTECHA, J. ZHANG ET. AL., Particle filtering, *IEEE Signal Processing Magazine*, **20** (2003), 19–38.

[19] D. DOLEV, J. HALPERN, AND H. R. STRONG, On the possibility and impossibility of achieving clock synchronization, in *Proceedings of the 16th Annual ACM Symposium on Theory of Computing, Washington DC, Apr–May 1984*, pp. 504–511. ACM, 1984.

[20] X. DU, M. GUIZANI, Y. XIAO, AND H. H. CHEN, Secure and efficient time synchronization in heterogeneous sensor networks, *IEEE Transactions on Vehicular Technology*, **57** (2008), 2387–2394.

[21] B. EFRON AND R. TIBSHIRANI, *An Introduction to the Bootstrap*, Chapman & Hall, 1993.

[22] J. ELSON, L. GIROD, AND D. ESTRIN, Fine-grained network time synchronization using reference broadcasts, in *Proceedings of the 5th Symposium on Operating System Design and Implementation, Boston, MA, December 2002*, pp. 147–163. ACM, 2002.

[23] J. ELSON AND K. ROMER, Wireless sensor networks: a new regime for time synchronization, in *Proceedings of the First Workshop on Hot Topics in Networks, Princeton, NJ, October 2002*, ACM, 2002.

[24] J. E. ELSON, Time synchronization in wireless sensor networks, Ph.D. dissertation, University of California Los Angeles, Los Angeles, CA, Apr 2003.

[25] A. EPHREMIDES, Energy concerns in wireless networks, *IEEE Transactions on Wireless Communications*, **9** (2002), 48–59.

[26] H. ERIKSSON, Mbone: the multicast backbone, *Communications of ACM*, **37** (1994), 54–60.

[27] N. FRERIS AND P. KUMAR, Fundamental limits on synchronization of affine clocks in networks, in *Proceedings of 46th IEEE Conference on Decision and Control, New Orleans, LA, December 2007*, pp. 921–926. IEEE, 2007.

[28] S. GANERIWAL, D. GANESAN, H. SHIM, V. TSIATSIS, AND M. B. SRIVASTAVA, Estimating clock uncertainty for efficient duty-cycling in sensor networks, in *Proceedings of the SenSys, San Diego, CA, November 2005*, pp. 130–141. ACM, 2005.

[29] S. GANERIWAL, R. KUMAR, AND M. SRIVASTAVA, Timing synch protocol for sensor networks, in *Proceedings of 1st International Conference on Embedded Network Sensor Systems, Los Angeles, CA, November 2003*, pp. 138–149. ACM, 2005.

[30] Q. GAO, K. J. BLOW, AND D. J. HOLDING, Simple algorithm for improving time synchronization in wireless sensor networks, *Electronics Letters*, **40** (2004), 889–891.

[31] J. GEHRKE, A. DOBRA, M. GAROFALAKIS, AND R. RASTOGI, Processing complex aggregate queries over data streams, in *Proceedings of the SIGMOD International Conference on Data Management, 2002*, pp. 61–72. ACM, 2002.

[32] J. V. GREUNEN AND J. RABAEY, Lightweight time synchronization for sensor networks, in *Proceedings of the 2nd ACM International Conference on Wireless Sensor Networks and Applications (WSNA), September 2003*, pp. 11–19. ACM, 2003.

[33] J. HEIDEMANN, W. YE, J. WILLS, A. SYED, AND Y. LI, Research challenges and applications for underwater sensor networking, in *Proceedings of the IEEE Wireless Communications and Networking Conference, Las Vegas, Nevada, April 2006*, pp. 1–5. `http://www.isi.edu/~johnh/PAPERS/Heidemann06a.html`. IEEE, 2006.

[34] J. L. HILL AND D. E. CULLER, Mica: a wireless platform for deeply embedded networks, *IEEE Micro*, **22** (2002), 12–24.

[35] J. P. HOBERT AND G. CASELLA, Functional compatibility, Markov chains, and Gibbs sampling with improper posteriors, *Journal Comp. Graph. Stat.*, **7** (1998), 42–60.

[36] Y.-W. HONG AND A. SCAGLIONE, A scalable synchronization protocol for large scale sensor networks and its applications, *IEEE Journal on Selected Areas in Communications*, **23** (2005), 1085–1099.

[37] M. HORAUER, U. SCHMID, K. SCHOSSMAIER, R. HOLLER, AND N. KERO, Psynutc-evaluation of a high precision time synchronization prototype system for ethernet lans, in *34th Annual Precise Time and Time Interval Meeting (PTTI), Dec 2002*. IEEE, 2002.

[38] D. A. HOWE, D. W. ALAN, AND J. A. BARNES, Properties of signal sources and measurements methods, in *Proceedings of the 35th Annual Symposium on Frequency Control, Philadelphia, Pennsylvania, May 1981*, pp. 669–716. ACM, 1981.

[39] A.-S. HU AND S. D. SERVETTO, Asymptotically optimal time synchronization in dense sensor networks, in *Proceedings of the 2nd ACM International Conference on Wireless Sensor Networks and Applications (WSNA), San Diego, CA, September 2003*, pp. 1–10. ACM, 2003.

[40] ——, On the scalability of cooperative time synchronization in pulse-connected networks, *IEEE Transactions on Information Theory*, **52** (2006), 2725–2748.

[41] D. R. JESKE, On the maximum likelihood estimation of clock offset, *IEEE Transactions on Communications*, **53** (2005), 53–54.

[42] N. JOHNSON, S. KOTZ, AND N. BALAKRISHNAN, *Continuous Univariate Distributions*, Vol. 1, 2nd edn. Wiley, 1994.

[43] S. J. JULIER AND J. K. UHLMANN, A general method for approximating non-linear transformations of probability distributions, Technical Report, Dept. of Engineering Science, University of Oxford, Nov 1996.

[44] S. M. KAY, *Fundamentals of Statistical Signal Processing*, Vol. I. *Estimation Theory*, Prentice Hall, 1993.

[45] H. KOPETZ AND W. OCHSENREITER, Clock synchronization in distributed real-time systems, *IEEE Transactions on Computers*, **36** (1987), 933–939.

[46] H. KOPETZ AND W. SCHWABL, Global time in distributed real-time systems, Technical Report 15/89, Technische Universität Wien, 1989.

[47] J. H. KOTECHA AND P. M. DJURIC, Gaussian particle filtering, *IEEE Transactions on Signal Processing*, **51** (2003), 2593–2602.

[48] ——, Gaussian sum particle filtering, *IEEE Transactions on Signal Processing*, **51** (2003), 2602–2612.

[49] S. KUNNIYUR AND S. NARASIMHAN, Modeling the effect of network parameters on delay in wireless ad hoc networks, in *Proceedings of the IEEE Conference on Sensor and Ad Hoc Communications and Networks, Santa Clara, CA, September 2005*, pp. 341–349. IEEE, 2005.

[50] E. L. LEHMANN AND G. CASELLA, *Theory of Point Estimation*, Springer, 1998.

[51] M. LEMMON, J. GANGULY, AND L. XIA, Model-based clock synchronization in networks with drifting clocks, in *Proceedings of the 2000 Pacific Rim International Symposium on Dependable Computing, Los Angeles, CA, December 2000*, pp. 177–185. IEEE, 2000.

[52] A. LEON-GARCIA, *Probability and Random Processes for Electrical Engineering*, 2nd edn. Addison-Wesley, 1993.

[53] Q. LI AND D. RUS, Global clock synchronization in sensor networks, *IEEE Transactions on Computers*, **55** (2006), 214–226.

[54] J. LIU, *Monte Carlo Strategies in Scientific Computing*, Springer, 2003.

[55] E. H. LLOYD, Least-squares estimation of location and scale parameters using order statistics, *Biometrika*, **39** (1952), 88–95.

[56] J. LUNDELIUS AND N. LYNCH, An upper and lower bound for clock synchronization, *Information and Control*, **62** (1984), 190–204.

[57] N. R. MANN, Optimum estimators for linear functions of location and scale parameters, *The Annals of Mathematical Statistics*, **40** (1969).

[58] M. MAROTI, B. KUSY, G. SIMON, AND A. LEDECZI, The flooding time synchronization protocol, in *Proceedings of the 2nd International Conference on Embedded Networked Sensor Systems, Baltimore, MD, November 2004*, pp. 39–49. ACM, 2004.

[59] K. MARZULLO AND S. OWICKI, Maintaining time in a distributed server, *Operating Systems Review*, **19** (1985).

[60] MICA2 AND MICA2DOT. `http://www.xbow.com/`.

[61] D. MILLS, Internet time synchronization: the network time protocol, *IEEE Transactions on Communications*, **39** (1991), 1482–1493.

[62] ——, *Computer Network Time Synchronization: The Network Time Protocol*, CRC Press, 2006.

[63] M. MOCK, R. FRINGS, E. NETT, AND S. TRIKALIOTIS, Continuous clock synchronization in wireless real-time applications, in *Proceedings of 19th IEEE Symposium on Reliable Distributed Systems (SRDS), October 2000*, p. 125. IEEE, 2000.

[64] S. MOON, P. SKELLEY, AND D. TOWSLEY, Estimation and removal of clock skew from network delay measurements, in *Proceedings of the IEEE INFOCOM Conference on Computer Communications, New York, NY, March 1999*, pp. 227–234. IEEE, 1999.

[65] K. SUN, P. NING, AND C. WONG, Secure and resilient clock synchronization in wireless sensor networks, *IEEE Journal on Selected Areas in Communications*, **24** (2006), 395–408.

[66] K. NOH, E. SERPEDIN, AND K. QARAQE, A new approach for time synchronization in wireless sensor networks: pairwise broadcast synchronization, *IEEE Transactions on Wireless Communications*, **7** (2008), 3318–3322.

[67] K.-L. NOH, Q. CHAUDHARI, E. SERPEDIN, AND B. SUTER, Novel clock phase offset and skew estimation using two-way timing message exchanges for wireless sensor networks, *IEEE Transactions on Communications*, **55** (2007), 766–777.

[68] K.-L. NOH AND E. SERPEDIN, Pairwise broadcast synchronization for wireless sensor networks, in *Proceedings of the IEEE International Workshop: From Theory To Practice in Wireless Sensor Networks, Helsinki, Finland, June* 2007, pp. 1–6. IEEE, 2007.

[69] S. PALCHAUDHURI, A. SAHA, AND D. B. JOHNSON, Adaptive clock synchronization in sensor networks, in *Proceedings of the Information Processing in Sensor Networks (IPSN), Berkeley, CA, April 2004*, pp. 340–348. ACM, 2004.

[70] A. PAPOULIS, *Probability, Random Variables and Stochastic Processes*, 3rd edn. McGraw-Hill, 1991.

[71] V. PAXSON, On calibrating measurements of packet transit times, in *Proceedings of the 7th ACM Sigmetrics Conference, Madison, WS, 1998*, Vol. 26, pp. 11–21. ACM, 1998.

[72] ——, On calibrating measurements of packet transit times, Tech. Report LBNL-41535, Lawrence Berkeley National Laboratory, March 1998.

[73] F. PERNKOPF AND D. BOUCHAFFRA, Genetic-based em algorithm for learning Gaussian mixture models, *IEEE Trans. On Pattern Analysis and Machine Intelligence*, **27** (2005), 1344–1348.

[74] S. PING, Delay measurement time synchronization for wireless sensor networks, Technical Report IRB-TR-03-013, Intel Research, Jun 2003.

[75] G. POTTIE AND W. KAISER, Wireless integrated network sensors, *Communications of the ACM*, **43** (2000), 51–58.

[76] R. RAMJEE, J. KUROSE, D. TOWSLEY, AND H. SCHULZRINNE, Adaptive playout mechanisms for packetized audio applications in wide-area networks, in *Proceeding of INFOCOM, Toronto, Canada, April 1994*, Vol. 2, pp. 680–688. IEEE, 1994.

[77] R. A. REDNER AND H. F. WALKER, Mixture densities, maximum likelihood and the em algorithm, *SIAM Review*, **26** (1984), 195–239.

[78] C. ROBERT AND G. CASELLA, *Monte Carlo Statistical Methods*, Springer, 1999.

[79] K. ROMER, Time synchronization in ad hoc networks, in *Proceedings of ACM Symposium on Mobile Ad Hoc Networking and Computing, Long Beach, CA, October 2001*, pp. 173–182. ACM, 2001.

[80] K. ROMER, P. BLUM, AND L. MEIER, Time synchronization and calibration in wireless sensor networks, in *Handbook of Sensor Networks: Algorithms and Architectures*, John Wiley & Sons, 2005.

[81] B. M. SADLER, Critical issues in energy-constrained sensor networks: synchronization, scheduling, and acquisition, in *Proceedings of IEEE International Conference on Acoustics, Speech, and Signal Processing, Philadelphia, PA*, Mar 2005, Vol. 5, pp. 785–788. IEEE, 2005.

[82] ——, Fundamentals of energy-constrained sensor network systems, *IEEE Aerospace Electrical Systems Magazine (Tutorials II)*, **20** (2005), 17–35.

[83] ——, Local and broadcast clock synchronization in a sensor node, *IEEE Signal Processing Letters*, **13** (2006), 9–12.

[84] B. M. SADLER AND A. SWAMI, Synchronization in sensor networks: an overview, in *Proceedings of IEEE Military Communications Conference, Washington DC, 2006*, pp. 1–6. IEEE, 2006.

[85] I. Sari, E. Serpedin, and B. Suter, On the joint synchronization of clock offset and skew in rbs-protocol, in *Proceedings of the Military Communications Conference, Washington DC, Oct 2006*. IEEE, 2006.

[86] M. L. Sichitiu and C. Veerarittiphan, Simple, accurate time synchronization for wireless sensor networks, in *Proceedings of IEEE Wireless Communications and Networking Conference (WCNC), New Orleans, LA, March 2003*, pp. 1266–1273. IEEE, 2003.

[87] G. Simon, M. Maroti, A. Ledeczi et al. Sensor network-based counter sniper system, in *Proceedings of the Second ACM Conference on Embedded Networked Sensor Systems (SenSys), November 2004*. ACM, 2004.

[88] F. Sivrikaya and B. Yener, Time synchronization in sensor networks: a survey, *IEEE Networks*, (2004), 45–50.

[89] W. Su, *Embedded Systems*, CRC Press, 2005.

[90] W. Su and I. F. Akyildiz, Time-diffusion synchronization protocol for wireless sensor networks, *IEEE/ACM Transactions on Networking*, **13** (2005), 384–397.

[91] B. Sundararaman, U. Buy, and A. D. Kshemkalyani, Clock synchronization for wireless sensor networks: a survey, *Ad-Hoc Networks*, 3 (2005), 281–323.

[92] A. Syed and J. Heidemann, Time synchronization for high latency acoustic networks, Technical Report ISI-TR-2005-602, USC/Information Sciences Institute, Apr 2005. `http://www.isi.edu/~johnh/PAPERS/Syed06a.html`.

[93] ——, Time synchronization for high latency acoustic networks, in *Proceedings of INFOCOM, Barcelona, Spain, April 2006*, pp. 1–12. IEEE, 2006.

[94] Z. Tian, X. Luo, and G. B. Giannakis, Cross-layer sensor network synchronization, in *Proceedings of the 38th Asilomar Conference on Signals, Systems and Computers, Pacific Grove, CA, November 2004*, Vol. 1, pp. 1276–1280. IEEE, 2004.

[95] D. Tulone, Resource-efficient time estimation for wireless sensor networks, in *Proceedings of the DIALM-POMC Workshop on Foundations of Mobile Computing, Philadelphia, PA, October 2004*, eds. S. Basagni and C. A. Philips, pp. 52–59. ACM, 2004.

[96] R. van der Merwe and E. Wan, Gaussian mixture sigma-point particle filters for sequential probabilistic inference in dynamic state-space models, in *Proceedings of the International Conference on Acoustics, Speech, and Signal Processing (ICASSP), April 2003*. IEEE, 2003.

[97] P. Veríssimo, L. Rodrigues, and A. Casimiro, CesiumSpray: a precise and accurate global time service for large-scale systems, Technical Report NAV-TR-97-0001, Universidade de Lisboa, 1997.

[98] J. R. Vig, Introduction to quartz frequency standards, Technical Report SLCET-TR-92-1, Army Research Laboratory Electronics and Power Sources Directorate, Oct 1992.

[99] G. Xing, C. Lu, Y. Zhang, Q. Huang, and R. Pless, Minimum power configuration in wireless sensor networks, in *Proceedings of the ACM MobiHoc, Urbana-Champaign, IL, May 2005*, pp. 390–401. ACM, 2005.

[100] F. Zhao and L. Guibas, *Wireless Sensor Networks: An Information Processing Approach*, Elsevier Inc., 2004.

[101] A. Zoubir and D. Iskander, *Bootstrap Techniques for Signal Processing*, Cambridge University Press, 2004.

INDEX